Herbert Spencer

Recent Discussions in Science, Philosophy and Morals

Herbert Spencer

Recent Discussions in Science, Philosophy and Morals

ISBN/EAN: 9783337033989

Printed in Europe, USA, Canada, Australia, Japan

Cover: Foto ©Thomas Meinert / pixelio.de

More available books at **www.hansebooks.com**

RECENT DISCUSSIONS

IN

SCIENCE, PHILOSOPHY, AND MORALS.

BY

HERBERT SPENCER,

AUTHOR OF "FIRST PRINCIPLES," "THE PRINCIPLES OF BIOLOGY," "THE
PRINCIPLES OF PSYCHOLOGY," ETC.

NEW YORK:

D. APPLETON AND COMPANY,

549 & 551 BROADWAY.

1871.

PREFACE.

THE present volume consists mainly of matter that is new to the American public. Three of the essays have not before appeared in this country, and two of the others, issued as a pamphlet, have had so small a circulation as to have been seen by but few readers. These several discussions have been drawn from Mr. Spencer at various times to correct misapprehensions and misrepresentations that have been made regarding the doctrines of his system of Philosophy. Some of them form valuable extensions of these doctrines, and all will be useful in promoting their right interpretation. Why the closing article has been taken from another volume and appended to this collection, requires a few words of explanation.

Seventeen years ago, Mr. Spencer published an elaborate Review article entitled "The Genesis of Science," in which he objected to Comte's views of the classification of the Sciences. Although Mr. Spencer's criticisms involved a radical dissent from the peculiar views of M. Comte, and what was held as fundamental in his philosophy, yet upon the publication of his own philosophical system Mr. Spencer found himself ranked as a positivist and a follower of Comte. Against this he repeatedly protested in public letters; but the charge was so continually reiterated that at length he found himself compelled to make a more formal statement of the differences between himself and the French philosopher. The result of this was a pamphlet published in 1864, in which he followed the rejection of Comte's classification by the promulgation of his own view, and appended a detailed statement of the differences between his doctrine and the doctrines of M. Comte. Some of his views of classification having been adversely criticised by Mr. Bain and Mr. Mill, he has replied to their strictures in a new article in the present volume. The general question is one of great interest to scientific students; and, for the convenience of those who desire to form an intelligent judgment of Mr. Spencer's case, both as contrasted with that of Comte, and on its own independent merits, it has been thought desirable to incorporate the original article on " The Genesis of Science " in this collection. Though placed last, it should be read first by those not already familiar with the discussion.

New York, *May*, 1871.

CONTENTS.

I.

MORALS AND MORAL SENTIMENTS.

[FROM THE FORTNIGHTLY REVIEW, APRIL, 1871.]

MORALS AND MORAL SENTIMENTS.

IF a writer who discusses unsettled questions takes up every gauntlet thrown down to him, polemical writing will absorb much of his energy. Having a power of work which unfortunately does not suffice for executing with any thing like due rapidity the task I have undertaken, I have made it a policy to avoid controversy as much as possible, even at the cost of being seriously misunderstood. Hence it happened that, when, in *Macmillan's Magazine* for July, 1869, Mr. Richard Hutton published, under the title of " A Questionable Parentage for Morals," a criticism upon a doctrine of mine, I decided to let his misrepresentations remain unnoticed until, in the course of my work, I arrived at the stage where, by a full exposition of this doctrine, they would be set aside. It did not occur to me that, in the mean time, these erroneous statements, accepted as true statements, would be repeated by other writers, and my views commented upon as untenable. This, however, has happened. In more periodicals than one, I have seen it asserted that Mr. Hutton has effectually disposed of my hypothesis. Supposing that this hypothesis has been rightly expressed by Mr. Hutton, Sir John Lubbock, in his " Origin of Civilization," etc., has been led to express a partial dissent ; which I think he would not have expressed had my own exposition been before him. Mr. Mivart, too, in his

recent " Genesis of Species," has been similarly betrayed
into misapprehensions. And now Sir Alexander Grant,
following the same lead, has conveyed to the readers of
the *Fortnightly Review* another of these conceptions,
which is but very partially true. Thus I find myself
compelled to say as much as will serve to prevent further
spread of the mischief.

If a general doctrine concerning a highly-involved
class of phenomena could be adequately presented in a
single paragraph of a letter, the writing of books would
be superfluous. In the brief exposition of certain ethical
doctrines held by me, which is given in Prof. Bain's
" Mental and Moral Science," it is stated that they are—

" as yet nowhere fully expressed. They form part of the more general doctrine of Evolution which he is engaged in working out ; and
they are at present to be gathered only from scattered passages. It
is true that, in his first work, ' Social Statics,' he presented what he
then regarded as a tolerably complete view of one division of Morals.
But, without abandoning this view, he now regards it as inadequate
—more especially in respect of its basis.''

Mr. Hutton, however, taking the bare enunciation of
one part of this basis, deals with it critically ; and, in the
absence of any exposition of it by me, sets forth what he
supposes to be my grounds for it, and proceeds to show
that they are unsatisfactory.

If, in his anxiety to suppress what he doubtless regards as a pernicious doctrine, Mr. Hutton could not wait
until I had explained myself, it might have been expected
that he would use whatever information was to be had
for rightly construing it. So far from seeking out such
information, however, he has, in a way for which I cannot account, ignored the information immediately before
him.

The title which Mr. Hutton has chosen for his criticism is, "A Questionable Parentage for Morals." Now, he has ample means of knowing that I allege a primary basis of Morals, quite independent of that which he describes and rejects. I do not refer merely to the fact that, having, when he reviewed "Social Statics,"[1] expressed his very decided dissent from this primary basis, he must have been aware that I allege it; for he may say that in the long interval which has elapsed he had forgotten all about it. But I refer to the distinct enunciation of this primary basis in that letter to Mr. Mill from which he quotes. In a preceding paragraph of the letter, I have explained that, while I accept utilitarianism in the abstract, I do not accept that current utilitarianism which recognizes for the guidance of conduct nothing beyond empirical generalizations; and I have contended that—

"Morality, properly so called—the science of right conduct—has for its object to determine *how* and *why* certain modes of conduct are detrimental, and certain other modes beneficial. These good and bad results cannot be accidental, but must be necessary consequences of the constitution of things; and I conceive it to be the business of Moral Science to deduce, from the laws of life and the conditions of existence, what kinds of action necessarily tend to produce happiness, and what kinds to produce unhappiness. Having done this, its deductions are to be recognized as laws of conduct; and are to be conformed to irrespective of a direct estimation of happiness or misery."

Nor is this the only enunciation of what I conceive to be the primary basis of morals, contained in this same letter. A subsequent paragraph, separated by four lines only from that which Mr. Hutton extracts, commences thus:

"Progressing civilization, which is of necessity a succession of compromises between old and new, requires a perpetual readjust-

[1] See *Prospective Review* for January, 1852.

ment of the compromise between the ideal and the practicable in social arrangements: to which end, both elements of the compromise must be kept in view. If it is true that pure rectitude prescribes a system of things far too good for men as they are, it is not less true that mere expediency does not of itself tend to establish a system of things any better than that which exists. While absolute morality owes to expediency the checks which prevent it from rushing into Utopian absurdities, expediency is indebted to absolute morality for all stimulus to improvement. Granted that we are chiefly interested in ascertaining what is *relatively right*, it still follows that we must first consider what is *absolutely right;* since the one conception presupposes the other."

I do not see how there could well be a more emphatic assertion that there exists a primary basis of morals independent of, and in a sense antecedent to, that which is furnished by experiences of utility; and, consequently, independent of, and in a sense antecedent to, those moral sentiments which I conceive to be generated by such experiences. Yet no one could gather from Mr. Hutton's article that I assert this; or would even find reasons for a faint suspicion that I do so. From the reference made to my further views, he would infer my acceptance of that empirical utilitarianism which I have expressly repudiated. And the title which Mr. Hutton gives to his paper clearly asserts, by implication, that I recognize no " parentage for morals " beyond that of the accumulation and organization of the effects of experience. I cannot believe that Mr. Hutton intended to convey this erroneous impression. He was, I suppose, too much absorbed in contemplating the proposition he combats to observe, or, at least, to attach any weight to, the propositions which accompany it. But I regret that he did not perceive the mischief he was likely to do me by spreading this one-sided statement.

I pass now to the particular question at issue—not

the " parentage for morals," but the parentage of moral
sentiments. In his version of my view on this more special doctrine, Mr. Hutton has similarly, I regret to say,
neglected the data which would have helped him to draw
an approximately true outline of it. It cannot well be
that the existence of such data was unknown to him.
They are contained in the " Principles of Psychology ; "
and Mr. Hutton reviewed that work when it was first
published.[1] In the chapter on The Feelings, which occurs
near the end of that work, there is sketched out a process of genesis by no means like that which Mr. Hutton
indicates ; and had he turned to that chapter he would
have seen that his description of the genesis of the moral
sentiments out of organized experiences is not such a one
as I should have given. Let me quote a passage from
that chapter :

" Not only are those emotions which form the immediate stimuli
to actions thus explicable, but the like explanation applies to the
emotions that leave the subject of them comparatively passive: as,
for instance, the emotion produced by beautiful scenery. The gradually increasing complexity in the groups of sensations and ideas co-
ordinated, ends in the coördination of those vast aggregations of
them which a grand landscape excites and suggests. The infant
taken into the midst of mountains is totally unaffected by them ;
but is delighted with the small group of attributes and relations presented in a toy. The child can appreciate, and be pleased with, the
more complicated relations of household objects and localities, the
garden, the field, and the street. But it is only in youth and mature
age, when individual things and small assemblages of them have
become familiar and automatically cognizable, that those immense
assemblages which landscapes present can be adequately grasped,
and the highly aggregated states of consciousness produced by them,
experienced. Then, however, the various minor groups of states,
that have been in earlier days severally produced by trees, by fields,

---

[1] His criticism will be found in the *National Review* for January, 1856,
under the title " Atheism."

by streams, by cascades, by rocks, by precipices, by mountains, by clouds, are aroused together. Along with the sensations immediately received, there are partially excited the myriads of sensations that have been in times past received from objects such as those presented; further, there are partially excited the various incidental feelings that were experienced on all these countless past occasions; and there are probably also excited certain deeper, but now vague, combinations of states, that were organized in the race during barbarous times, when its pleasurable activities were chiefly among the woods and waters. And out of all these excitations, some of them actual, but most of them nascent, is composed the emotion which a fine landscape produces in us."

It is, I think, amply manifest that the processes here indicated are not to be taken as intellectual processes— not as processes in which recognized relations between pleasures and their antecedents, or intelligent adaptations of means to ends, form the dominant elements. The state of mind produced by an aggregate of picturesque objects is not one resolvable into propositions. The sentiment does not contain within itself any consciousness of causes and consequences of happiness. The vague recollections of other beautiful scenes and other delightful days which it dimly rouses, are not aroused because of any rational coördinations of ideas that have been formed in by-gone days. Mr. Hutton, however, has assumed that in the genesis of moral feelings as due to inherited experiences of the pleasures and pains arising from certain modes of conduct, I am speaking of reasoned-out experiences— experiences consciously accumulated and generalized. He altogether overlooks the fact that the genesis of emotions is distinguished from the genesis of ideas in this: that whereas the ideas are composed of elements that are simple, definitely related, and (in the case of general ideas) constantly related, emotions are composed of enormously complex aggregates of elements which are never twice alike, and that stand in relations which are never twice alike. The difference in the resulting modes of consciousness is this: In the genesis of an idea the successive experiences, be they of sounds, colors, touches, tastes, or be they of the special objects that combine many of these into groups, have so much in common that each, when it occurs, can be definitely thought of as like those which preceded it. But in the genesis of an emotion the successive experiences so far differ that each of them, when it occurs, suggests past experiences which are not specifically similar, but have only a general similarity; and, at the same time, it suggests benefits or evils in past experience which likewise are various in their special natures, though they have a certain community of general nature. Hence it results that the consciousness aroused is a multitudinous, confused consciousness, in which, along with a certain kind of combination among the impressions received from without, there is a vague cloud of ideal combinations akin to them, and a vague mass of ideal feelings of pleasure or pain that were associated with these. We have abundant proof that feelings grow up without reference to recognized causes and consequences, and without the possessor of them being able to say why they have grown up; though analysis, nevertheless, shows that they have been formed out of connected experiences. The familiar fact to which, I suppose, almost every one can testify, that a kind of jam which was, during childhood, repeatedly taken after medicine, may become by simple association of sensations so nauseous that it cannot be tolerated in after-life, illustrates clearly enough the way in which repugnances may be established by habitual association of feelings, without any idea of causal connection; or rather, in spite of the knowledge that there is no causal connection. Similarly with pleasurable emotions.

The cawing of a rook is not in itself an agreeable sound
—musically considered, it is very much the contrary.
Yet the cawing of rooks usually produces in people very
pleasurable feelings—feelings which most of them suppose
to result from the quality of the sound itself. Only the
few who are given to self-analysis are aware that the
cawing of rooks is agreeable to them because it has been
connected with countless of their greatest gratifications—
with the gathering of wild-flowers in childhood; with
Saturday-afternoon excursions in school-boy days; with
midsummer holidays in the country, when books were
thrown aside, and lessons were replaced by games and
adventures in the fields; with fresh, sunny mornings in
after-years, when a walking-excursion was an immense
relief from toil. As it is, this sound, though not causally
related to all these multitudinous and varied past delights,
but only often associated with them, can no more be
heard without rousing a dim consciousness of these delights, than the voice of an old friend unexpectedly coming
into the house can be heard without suddenly raising a
wave of that feeling that has resulted from the pleasures
of past companionship. If we are to understand the
genesis of emotions, either in the individual or in the
race, we must take account of this all-important process.
Mr. Hutton, however, apparently overlooking it, and not
having reminded himself, by referring to the " Principles
of Psychology," that I insist upon it, represents my hypothesis to be that a certain sentiment results from the
consolidation of intellectual conclusions! He speaks of
me as believing that "what seems to us now the ' necessary' intuitions and _a priori_ assumptions of human
nature, are likely to prove, when scientifically analyzed,
nothing but a similar conglomeration of our ancestors'
best observations and most useful empirical rules." He

supposes me to think that men having, in past times, come to *see* that truthfulness was useful, "the habit of approving truth-speaking and fidelity to engagements, which was first based on this ground of utility, became so rooted, that the utilitarian ground of it was forgotten, and *we* find ourselves springing to the belief in truth-speaking and fidelity to engagements from an inherited tendency." Similarly throughout, Mr. Hutton has so used the word "utility," and so interpreted it on my behalf, as to make me appear to mean that moral sentiment is formed out of *conscious generalizations* respecting what is beneficial and what detrimental. Were such my hypothesis, his criticisms would be very much to the point; but as such is not my hypothesis, they fall to the ground. The experiences of utility I refer to are those which become registered, not as distinctly-recognized connections between certain kinds of acts and certain kinds of remote results, but those which become registered in the shape of associations between groups of feelings that have often recurred together, though the relation between them has not been consciously generalized—associations the origin of which may be as little perceived as is the origin of the pleasure given by the sounds of a rookery; but which, nevertheless, have arisen in the course of daily converse with things, and serve as incentives or deterrents.

In the paragraph which Mr. Hutton has extracted from my letter to Mr. Mill, I have indicated an analogy between those effects of emotional experiences out of which I believe moral sentiments have been developed, and those effects of intellectual experiences out of which I believe space-intuitions have been developed. Rightly considering that the first of these hypotheses cannot stand if the last is disproved, Mr. Hutton has directed part of

his attack against this last.   But would it not have been
well if he had referred to the " Principles of Psychology,"
where this last hypothesis is set forth at length, before
criticising it?   Would it not have been well to have
given an abstract of my own description of the process,
instead of substituting what he *supposes* my description
must be?   Any one who turns to the " Principles of Psychology " (first edition, pp. 218–245), and reads the two
chapters, The Perception of Body as presenting Statical
Attributes, and The Perception of Space, will find that
Mr. Hutton's account of my view on this matter has
given him no notion of the view as it is expressed by me ;
and will, perhaps, be less inclined to smile than he was
when he read Mr. Hutton's account.   I cannot here do
more than thus imply the invalidity of such part of Mr.
Hutton's argument as proceeds upon this incorrect representation.   The pages that would be required for properly
explaining the doctrine that space-intuitions result from
organized experiences may be better used for explaining
this analogous doctrine at present before us.   This I will
now endeavor to do ; not indirectly by correcting misapprehensions, but directly by an exposition which shall be
as brief as the extremely involved nature of the process
allows.

An infant in arms, that is old enough to gaze at
objects around with some vague recognition, smiles in
response to the laughing face and soft, caressing voice of
its mother.   Let there come some one who, with an angry
face, speaks to it in loud, harsh tones.   The smile disappears, the features contract into an expression of pain,
and, beginning to cry, it turns away its head and makes
such movements of escape as are possible.   What is the
meaning of these facts?   Why does not the frown make
it smile, and the mother's laugh make it weep?   There

is but one answer. Already in its developing brain there is coming into play the structure through which one cluster of visual and auditory impressions excites pleasurable feelings, and the structure through which another cluster of visual and auditory impressions excites painful feelings. The infant knows no more about the relation existing between a ferocious expression of face, and the evils that may follow the perception of it, than the young bird just out of its nest knows of the possible pain and death which may be inflicted by a man coming toward it; and as certainly in the one case as in the other, the alarm felt is due to a partially-established nervous structure. Why does this partially-established nervous structure betray its presence thus early in the human being? Simply because, in the past experiences of the human race, smiles and gentle tones in those around have been the habitual accompaniments of pleasurable feelings; while pains of many kinds, immediate and more or less remote, have been continually associated with the impressions received from knit brows and set teeth and grating voice. Much deeper down than the history of the human race must we go to find the beginnings of these connections. The appearances and sounds which excite in the infant a vague dread, indicate danger; and do so because they are the physiological accompaniments of destructive action—some of them common to man and inferior mammals, and consequently understood by inferior mammals, as every puppy shows us. What we call the natural language of anger, is due to a partial contraction of those muscles which actual combat would call into play; and all marks of irritation, down to that passing shade over the brow which accompanies slight annoyance, are incipient stages of these same contractions. Conversely with the natural language of pleasure, and of that state of mind which we call amicable feeling : this, too, has a physiological interpretation.[1]

Let us pass now from the infant in arms to the children in the nursery. What have the experiences of each one of these been doing in aid of the emotional development we are considering? While its limbs have been growing more agile by exercise, its manipulative skill increasing by practice, its perceptions of objects growing by use quicker, more accurate, more comprehensive; the associations between these two sets of impressions received from those around, and the pleasures and pains received along with them, or after them, have been by frequent repetition made stronger, and their adjustments better. The dim sense of pain and the vague glow of delight which the infant felt, have, in the urchin, severally taken shapes that are more definite. The angry voice of a nurse-maid no longer arouses only a formless feeling of dread, but also a specific idea of the slap that may follow. The frown on the face of a bigger brother, along with the primitive, indefinable sense of ill, brings the sense of ills that are definable in thought as kicks, and cuffs, and pullings of hair, and losses of toys. The faces of parents, looking now sunny, now gloomy, have grown to be respectively associated with multitudinous forms of gratification and multitudinous forms of discomfort or privation. Hence these appearances and sounds, which imply amity or enmity in those around, become symbolic of happiness and misery ; so that eventually perception of the one set or the other can scarcely occur without raising a wave of pleasurable feeling or of painful feeling. The body of this wave is still substantially of the same nature as it was

---

[1] Hereafter I hope to elucidate at length these phenomena of expression. For the present, I can refer only to such further indications as are contained in two essays on The Physiology of Laughter and the Origin and Function of Music.

at first; for though in each of these multitudinous experiences a special set of facial and vocal signs has been connected with a special set of pleasures or pains, yet since these pleasures or pains have been immensely varied in their kinds and combinations, and since the signs that preceded them were in no two cases quite alike, it results that to the last the consciousness produced remains as vague as it is voluminous. The myriads of partially-aroused ideas resulting from past experiences are massed together and superposed, so as to form an aggregate in which nothing is distinct, but which has the character of being pleasurable or painful according to the nature of its original components ; the chief difference between this developed feeling and the feeling aroused in the infant being, that on bright or dark background forming the body of it, may now be sketched out in thought the particular pleasures or pains which the particular circumstances suggest as likely.

What must be the working of this process under the conditions of aboriginal life? The emotions given to the young savage by the natural language of love and hate in the members of his tribe, gain first a partial definiteness in respect to his intercourse with his family and play-mates ; and he learns by experience the utility, in so far as his own ends are concerned, of avoiding courses which call from others manifestations of anger, and taking courses which call from them manifestations of pleasure. Not that he consciously generalizes. He does not at that age, probably not at any age, formulate his experiences in the general principle that it is well for him to do things which bring smiles from others, and to avoid doing things which bring frowns. What happens is, that having, in the way shown, inherited this connection between the perception of anger in others and the feeling of dread, and having

discovered that particular acts of his bring on this anger,
he cannot subsequently think of committing one of these
acts without thinking of the resulting anger, and feeling
more or less of the resulting dread. He has no thought
of the utility or inutility of the act itself; the deterrent is
the mainly vague, but partially definite, fear of evil that
may follow. So understood, the deterring emotion is
one that has grown out of experiences of utility, using that
word in its ethical sense; and if we ask why this dreaded
anger is called forth from others, we shall habitually find
that it is because the forbidden act entails pain somewhere—is negatived by utility. On passing from the
domestic injunctions to the injunctions current in the
tribe, we see no less clearly how these emotions produced
by approbation and reprobation come to be connected in
experience with actions that are beneficial to the tribe,
and actions that are detrimental to the tribe; and how
there consequently grow up incentives to the one class of
actions and prejudices against the other class. From early
boyhood the young savage hears recounted the daring
deeds of his chief—hears them in words of praise, and sees
all faces glowing with admiration. From time to time
also he listens while some one's cowardice is described
in tones of scorn, and with contemptuous metaphors, and
sees him meet with derision and insult whenever he
appears. That is to say, one of the things that comes to
be strongly associated in his mind with smiling faces,
which are symbolical of pleasures in general, is courage;
and one of the things that comes to be associated in his
mind with frowns and other marks of enmity, which form
his symbol of unhappiness, is cowardice. These feelings
are not formed in him because he has reasoned his way
to the truth that courage is useful to the tribe, and, by
implication, to himself, or to the truth that cowardice is a

cause of evil. In adult life he may, perhaps, see this; but he certainly does not see it at the time when bravery is thus associated in his consciousness with all that is good, and cowardice with all that is bad. Similarly there are produced in him feelings of inclination or repugnance toward other lines of conduct that have become established or interdicted, because they are beneficial or injurious to the tribe; though neither the young nor the adults know why they have become established or interdicted. Instance the praiseworthiness of wife-stealing, and the viciousness of marrying within the tribe.

We may now ascend a stage to an order of incentives and restraints derived from these. The primitive belief is that every dead man becomes a demon, who remains somewhere at hand, may at any moment return, may give aid or do mischief, and is continually propitiated. Hence, among other agents whose approbation or reprobation is contemplated by the savage as a consequence of his conduct, are the spirits of his ancestors. When a child he is told of their deeds, now in triumphant tones, now in whispers of horror; and the instilled belief that they may inflict some vaguely-imagined but fearful evil, or give some great help, becomes a powerful incentive or deterrent. Especially does this happen when the narrative is of a chief, distinguished for his strength, his ferocity, his persistence in that revenge which the experiences of the savage make him regard as beneficial and virtuous. The consciousness that such a chief, dreaded by neighboring tribes, and dreaded, too, by members of his own tribe, may reappear and punish those who have disregarded his injunctions, becomes a powerful motive. But it is clear, in the first place, that the imagined anger and the imagined satisfaction of this deified chief are simply transfigured forms of the anger and satisfaction displayed by those around; and

that the feelings accompanying such imaginations have the
same original root in the experiences which have associated
an average of painful results with the manifestation of
another's anger, and an average of pleasurable results with
the manifestation of another's satisfaction.   And it is
clear, in the second place, that the actions thus forbidden
and encouraged must be mostly actions that are respectively detrimental and beneficial to the tribe; since the
successful chief is usually a better judge than the rest,
and has the preservation of the tribe at heart.   Hence
experiences of utility, consciously or unconsciously organized, underlie his injunctions; and the sentiments which
prompt obedience are, though very indirectly and without
the knowledge of those who feel them, referable to experiences of utility.

This transfigured form of restraint, differing at first
but little from the original form, admits of immense
development.   Accumulating traditions, growing in
grandeur as they are repeated from generation to generation, make more and more superhuman the early-recorded
hero of the race.   His powers of inflicting punishment
and giving happiness become ever greater, more multitudinous and varied; so that the dread of divine dis-
pleasure, and the desire to obtain divine approbation,
acquire a certain largeness and generality.   Still the conceptions remain anthropomorphic.   The revengeful deity
continues to be thought of in terms of human emotions,
and continues to be represented as displaying these emotions in human ways.   Moreover, the sentiments of right
and duty, so far as they have become developed, refer
mainly to divine commands and interdicts; and have
little reference to the natures of the acts commanded or
interdicted.   In the intended offering up of Isaac, in the
sacrifice of Jephthah's daughter, and in the hewing to

EARLY MEANING OF RIGHT AND WRONG.   25

pieces of Agag, as much as in the countless atrocities committed from religious motives by other early historic
races, we see that the morality and immorality of actions,
as we understand them, are at first little recognized ; and
that the feelings, chiefly of dread, which serve in place of
them, are feelings felt toward the unseen beings supposed
to issue the commands and interdicts.

Here it will be said that, as just admitted, these are
not the moral sentiments properly so called. This is true.
They are simply sentiments that precede and make
possible those highest sentiments which do not refer either
to personal benefits or evils to be expected from men, or
to more remote rewards and punishments. Several comments are, however, called forth by this criticism. One
is, that if we glance back at past beliefs and their correlative feelings, as shown in Dante's poem, in the mystery-
plays of the middle ages, in St. Bartholomew massacres,
in burnings for heresy, we get proof that in comparatively
modern times right and wrong meant little else than sub-
ordination or insubordination—to a divine ruler primarily
and under him to a human ruler. Another is, that down
to our own day this conception largely prevails, and is
even embodied in elaborate ethical works—instance the
"Essays on the Principles of Morality," by Jonathan
Dymond, which recognizes no ground of moral obligation,
save the will of God as expressed in the current creed.
And yet a further is, that while in sermons the torments
of the damned and the joys of the blessed are set forth as
the dominant deterrents and incentives, and while we
have prepared for us printed instructions " how to make
the best of both worlds," it cannot be denied that the
feelings which impel and restrain men are still largely
composed of elements like those operative on the savage—
the dread, partly vague, partly specific, associated with

2

the idea of reprobation, human and divine, and the sense of satisfaction, partly vague, partly specific, associated with the idea of approbation, human and divine.

But during the growth of that civilization which has been made possible by these ego-altruistic sentiments, there have been slowly evolving the altruistic sentiments. Development of these has gone on only as fast as society has advanced to a state in which the activities are mainly peaceful. The root of all the altruistic sentiments is sympathy; and sympathy could become dominant only when the mode of life, instead of being one that habitually inflicted direct pain, became one which conferred direct and indirect benefits; the pains inflicted being mainly incidental and indirect. Adam Smith made a large step toward this truth when he recognized sympathy as giving rise to these superior controlling emotions. His "Theory of Moral Sentiments," however, requires to be supplemented in two ways. The natural process by which sympathy becomes developed into a more and more important element of human nature, has to be explained; and there has also to be explained the process by which sympathy produces the highest and most complex of the altruistic sentiments—that of justice. Respecting the first process, I can here do no more than say that sympathy may be proved, both inductively and deductively, to be the concomitant of gregariousness; the two having all along increased by reciprocal aid. Multiplication has ever tended to force into an association, more or less close, all creatures having kinds of food and supplies of food that permit association; and established psychological laws warrant the inference that some sympathy will inevitably result from habitual manifestations of feelings in presence of one another, and that the gregariousness being augmented by the increase of sympathy, further facilitates the development of sympathy. But there are negative and positive checks upon this development—negative, because sympathy cannot advance faster than intelligence advances, since it presupposes the power of interpreting the natural language of the various feelings, and of mentally representing those feelings; positive, because the immediate needs of self-preservation are often at variance with its promptings, as, for example, during the predatory stages of human progress. For explanations of the second process, I must refer to "The Principles of Psychology" (§ 202, first edition, and § 215, second edition) and to "Social Statics," Part II., Chapter V.[1] Asking that in default of space these explanations may be taken for granted, let me here point out in what sense even sympathy, and the sentiments that result from it, are due to experiences of utility. If we suppose all thought of rewards or punishments, immediate or remote, to be left out of consideration, it is clear that any one who hesitates to inflict a pain because of the vivid representation of that pain which rises in his consciousness, is restrained, not by any sense of obligation or by any formulated doctrine of utility, but by the painful association established in him. And it is clear that if, after repeated experiences of the moral discomfort he has felt from witnessing the unhappiness indirectly caused by some of his acts, he is led to check himself when again tempted to those acts, the restraint is of like nature. Conversely with the pleasure-giving acts: repetitions of kind deeds, and experiences of the sympathetic gratifications that follow, tend continually to make stronger the association between such deeds and feelings of happiness.

---

[1] I may add that in "Social Statics," Chapter XXX., I have indicated, in a general way, the causes of the development of sympathy and the restraints upon its development—confining the discussion, however, to the case of the human race, my subject limiting me to that. The accompanying teleology I now disclaim.

Eventually these experiences may be consciously generalized, and there may result a deliberate pursuit of the
sympathetic gratifications.   There may also come to be
distinctly recognized the truths that the remoter results
are respectively detrimental and beneficial — that due
regard for others is conducive to ultimate personal welfare
and disregard of others to ultimate personal disaster; and
then there may become current such summations of experience as " honesty is the best policy."   But so far from
regarding these intellectual recognitions of utility as
preceding and causing the moral sentiment, I regard the
moral sentiment as preceding such recognitions of utility,
and making them possible.   The pleasures and pains
directly resulting in experience from sympathetic and
unsympathetic actions, had first to be slowly associated
with such actions, and the resulting incentives and deterrents frequently obeyed, before there could arise the
perceptions that sympathetic and unsympathetic actions are remotely beneficial or detrimental to the actor;
and they had to be obeyed still longer and more generally before there could arise the perceptions that they
are socially beneficial or detrimental.   When, however,
the remote effects, personal and social, have gained
general recognition, are expressed in current maxims,
and lead to injunctions having the religious sanction,
the sentiments that prompt sympathetic actions and
check unsympathetic ones are immensely strengthened
by their alliances.   Approbation and reprobation, divine
and human, come to be associated in thought with
the sympathetic and unsympathetic actions respectively.
The commands of the creed, the legal penalties, and
the code of social conduct, unitedly enforce them;
and every child as it grows up, daily has impressed
on it, by the words and faces and voices of those around,

the authority of these highest principles of conduct. And now we may see why there arises a belief in the special sacredness of these highest principles, and a sense of the supreme authority of the altruistic sentiments answering to them. Many of the actions which, in early social states, received the religious sanction and gained public approbation, had the drawback that such sympathies as existed were outraged, and there was hence an imperfect satisfaction. Whereas these altruistic actions, while similarly having the religious sanction and gaining public approbation, bring a sympathetic consciousness of pleasure given or of pain prevented; and beyond this, bring a sympathetic consciousness of human welfare at large, as being furthered by making altruistic actions habitual. Both this special and this general sympathetic consciousness become stronger and wider in proportion as the power of mental representation increases, and the imagination of consequences, immediate and remote, grows more vivid and comprehensive. Until at length these altruistic sentiments begin to call in question the authority of those ego-altruistic sentiments which once ruled unchallenged. They prompt resistance to laws that do not fulfil the conception of justice, encourage men to brave the frowns of their fellows by pursuing a course at variance with customs that are perceived to be socially injurious, and even cause dissent from the current religion; either to the extent of disbelief in those alleged divine attributes and acts not approved by this supreme moral arbiter, or to the extent of entire rejection of a creed which ascribes such attributes and acts.

Much that is required to make this hypothesis complete must stand over until, at the close of the second volume of "The Principles of Psychology," I have space for a full exposition. What I have said will make it sufficiently clear that two fundamental errors have been made in the interpretation put upon it. Both Utility and Experience have been construed in senses much too narrow. Utility, convenient a word as it is from its comprehensiveness, has very inconvenient and misleading implications. It vividly suggests uses and means and proximate ends, but very faintly suggests the pleasures, positive or negative, which are the ultimate ends, and which, in the ethical meaning of the word, are alone considered; and, further, it implies conscious recognition of means and ends—implies the deliberate taking of some course to gain a perceived benefit. Experience, too, in its ordinary acceptation, connotes definite perceptions of causes and consequences, as standing in observed relations, and is not taken to include the connections formed in consciousness between states that recur together, when the relation between them, causal or other, is not perceived. It is in their widest senses, however, that I habitually use these words, as will be manifest to every one who reads the " Principles of Psychology ; " and it is in these widest senses that I have used them in the letter to Mr. Mill. I think I have shown above that, when they are so understood, the hypothesis briefly set forth in that letter is by no means so indefensible as is supposed. At any rate, I have shown—what seemed for the present needful to show—that Mr. Hutton's versions of my views must not be accepted as correct.

HERBERT SPENCER.

II.

THE ORIGIN OF ANIMAL-WORSHIP.

[FROM THE FORTNIGHTLY REVIEW, MAY, 1870.]

THE ORIGIN OF ANIMAL-WORSHIP.

Mr. McLennan's recent essays on the Worship of Animals and Plants have done much to elucidate a very obscure subject. By pursuing in this case, as before in another case, the truly scientific method of comparing the phenomena presented by existing uncivilized races with those which the early traditions of civilized races present, he has rendered both more comprehensible than they were before.

It seems to me, however, that Mr. McLennan gives but an indefinite answer to the essential question—How did the worship of animals and plants arise? Indeed, in his concluding paper, he expressly leaves this problem without a solution; saying that his "is not an hypothesis explanatory of the origin of *Totemism*, be it remembered, but an hypothesis explanatory of the animal and plant worship of the ancient nations." So that we have still to ask—Why have savage tribes so generally taken animals and plants and other things as their totems? What can have induced this tribe to ascribe special sacredness to one creature, and that tribe to another? And if to these questions the general reply is, that each tribe considers itself to be descended from the object of its reverence, then there presses for answer the further question—How came so strange a notion into existence? If this notion occurred

in one case only, we might set it down to some whim of
thought or some illusive occurrence. But appearing as it
does with multitudinous variations among so many uncivilized races in different parts of the world, and having
left equally numerous traces in the superstitions of the
extinct civilized races, we cannot assume any special or
exceptional cause. Moreover, the general cause, whatever
it may be, must be such as does not negative an aboriginal
intelligence essentially like our own. After studying the
grotesque beliefs of savages, we are apt to suppose that
their reason is not as our reason. But this supposition is
inadmissible. Given the amount of knowledge which
primitive men possess, and given the imperfect verbal
symbols used by them in speech and thought, and the conclusions they habitually reach will be those that are *relatively* the most rational. This must be our postulate;
and, setting out with this postulate, we have to ask how
primitive men came so generally, if not universally, to believe themselves the progeny of animals or plants or inanimate bodies. There is, I believe, a satisfactory answer.

The proposition with which Mr. McLennan sets out,
that totem-worship preceded the worship of anthropomorphic gods, is one to which I can yield but a qualified as-
sent. It is true in a sense, but not wholly true. If the
words "gods" and "worship" carry with them their ordinary definite meanings, the statement is true; but if
their meanings are widened so as to comprehend those
earliest vague notions out of which the definite ideas of
gods and worship are evolved, I think it is not true. The
rudimentary form of all religion is the propitiation of dead
ancestors, who are supposed to be still existing, and to be
capable of working good or evil to their descendants. As
a preparation for dealing hereafter with the principles of

sociology, I have, for some years past, directed much attention to the modes of thought current in the simpler human societies; and evidence of many kinds, furnished by all varieties of uncivilized men, has forced on me a conclusion harmonizing with that lately expressed in this Review by Prof. Huxley—namely, that the savage, conceiving a corpse to be deserted by the active personality who dwelt in it, conceives this active personality to be still existing, and that his feelings and ideas concerning it form the basis of his superstitions. Everywhere we find expressed or implied the belief that each person is double; and that when he dies, his other self, whether remaining near at hand or gone far away, may return, and continues capable of injuring his enemies and aiding his friends.[1]

[1] A critical reader may raise an objection. If animal-worship is to be rationally interpreted, how can the interpretation set out by assuming a belief in the spirits of dead ancestors—a belief which just as much requires explanation? Doubtless there is here a wide gap in the argument. I hope eventually to fill it up. Here, out of many experiences which conspire to generate this belief, I can but briefly indicate the leading ones : 1. It is not impossible that his shadow, following him everywhere, and moving as he moves, may have some small share in giving to the savage a vague idea of his duality. It needs but to watch a child's interest in the movements of its shadow, and to remember that at first a shadow cannot be interpreted as a negation of light, but is looked upon as an entity, to perceive that the savage may very possibly consider it as a specific something which forms part of him. 2. A much more decided suggestion of the same kind is likely to result from the reflection of his face and figure in water: imitating him as it does in his form, colors, motions, grimaces. When we remember that not unfrequently a savage objects to have his portrait taken, because he thinks whoever carries away a representation of him carries away some part of his being, will see how probable it is that he thinks his double in the water is a reality in some way belonging to him. 3. Echoes must greatly tend to confirm the idea of duality otherwise arrived at. Incapable as he is of understanding their natural origin, the primitive man necessarily ascribes them to living beings—beings who mock him and elude his search. 4. The suggestions resulting from these and other physical phenomena are, however, secondary in importance. The root of this belief in another self lies in the experience of dreams. The distinction so easily made by us between our life in dreams and our real life, is one which the savage recognizes in but a vague way; and he cannot express even that distinction which he perceives. When he awakes, and to those who have seen

But how out of the desire to propitiate this second personality of a deceased man (the words " ghost " or " spirit "
are somewhat misleading, since the savage believes that
the second personality reappears in a form equally tangible with the first) does there grow up the worship of

him lying quietly asleep, describes where he has been, and what he has done,
his rude language fails to state the difference between seeing and dreaming
that he saw, doing and dreaming that he did. From this inadequacy of his
language it not only results that he cannot truly represent this difference to
others, but also that he cannot truly represent it to himself. Hence, in the
absence of an alternative interpretation, his belief, and that of those to whom
he tells his adventures, is that his other self has been away and came back
when he awoke. And this belief, which we find among various existing savage tribes, we equally find in the traditions of the early civilized races.
5. The conception of another self capable of going away and returning, receives what to the savage must seem conclusive verifications from the abnor-
mal suspensions of consciousness, and derangements of consciousness, that
occasionally occur in members of his tribe. One who has fainted, and cannot
be immediately brought back to himself (note the significance of our own
phrases "returning to himself," etc.) as a sleeper can, shows him a state in
which the other self has been away for a time beyond recall. Still more is
this prolonged absence of the other self shown him in cases of apoplexy, catalepsy, and other forms of suspended animation. Here for hours the other
self persists in remaining away, and on returning refuses to say where he has
been. Further verification is afforded by every epileptic subject, into whose
body, during the absence of the other self, some enemy has entered; for how
else does it happen that the other self on returning denies all knowledge of
what his body has been doing? And this supposition that the body has been
"possessed" by some other being, is confirmed by the phenomena of som-
nambulism and insanity. 6. What, then, is the interpretation inevitably put
upon death? The other self has habitually returned after sleep, which simulates death. It has returned, too, after fainting, which simulates death much
more. It has even returned after the rigid state of catalepsy, which simulates
death very greatly. Will it not return also after this still more prolonged
quiescence and rigidity? Clearly it is quite possible—quite probable even.
The dead man's other self is gone away for a long time, but it still exists somewhere, far or near, and may at any moment come back to do all he said he
would do. Hence the various burial-rites—the placing of weapons and valu-
ables along with the body, the daily bringing of food to it, etc. I hope hereafter to show that, with such knowledge of the facts as he has, this interpretation is the most reasonable the savage can arrive at. Let me here, however,
by way of showing how clearly the facts bear out this view, give one illustration out of many. " The ceremonies with which they [the Veddahs] invoke
them [the shades of the dead] are few as they are simple. The most common

animals, plants, and inanimate objects? Very simply.
Savages habitually distinguish individuals by names that
are either directly suggestive of some personal trait or fact
of personal history, or else express an observed community
of character with some well-known object. Such a genesis of individual names, before surnames have arisen, is
inevitable; and how easily it arises we shall see on remembering that it still goes on in its original form, even
when no longer needful. I do not refer only to the significant fact that in some parts of England, as in the nail-
making districts, nicknames are universal, and surnames
scarcely recognized; but I refer to the general usage
among both children and adults. The rude man is apt to
be known as "a bear;" a sly fellow, as an "old fox;" a
hypocrite, as "the crocodile." Names of plants, too, are
used; as when the red-haired boy is called "carrots" by
his school-fellows. Nor do we lack nicknames derived
from inorganic objects and agents: instance that given by
Mr. Carlyle to the elder Sterling—"Captain Whirlwind."
Now, in the earliest savage state, this metaphorical nam-

is the following: An arrow is fixed upright in the ground, and the Veddah
dances slowly round it, chanting this invocation, which is almost musical in
its rhythm:

　　"Mâ miyâ, mâ miy, mâ deyâ,
　　　Topaug Koyichetti mittigan yandâh?"

　　"My departed one, my departed one, my God!
　　　Where art thou wandering?"

"This invocation appears to be used on all occasions when the intervention
of the guardian spirits is required in sickness, preparatory to hunting, etc.
Sometimes in the latter case, a portion of the flesh of the game is promised as
a votive offering, in the event of the chase being successful; and they believe
that the spirits will appear to them in dreams and tell them where to hunt.
Sometimes they cook food and place it in the dry bed of a river, or some other
secluded spot, and then call on their deceased ancestors by name, 'Come and
partake of this! Give us maintenance as you did when living! Come, where-
soever you may be, on a tree, on a rock, in the forest, come!' And dance
round the food, half chanting half shouting the invocation."—*Bailey, Trans.
Eth. Soc.*, London, N. S., ii., p. 301.

ing will in most cases commence afresh in each generation
—must do so, indeed, until surnames of some kind have
been established. I say in most cases, because there will
occur exceptions in the cases of men who have distinguished themselves. If "the Wolf," proving famous in
fight, becomes a terror to neighboring tribes, and a dominant man in his own, his sons, proud of their parentage,
will not let fall the fact that they descended from the
Wolf; nor will this fact be forgotten by the rest of the
tribe who held "the Wolf" in awe, and see some reason
to dread his sons. In proportion to the power and celeb-
rity of the Wolf will this pride and this fear conspire to
maintain among his grandchildren and great-grandchildren, as well as among those over whom they dominate,
the remembrance of the fact that their ancestor was the
Wolf. And if, as will occasionally happen, this dominant
family becomes the root of a new tribe, the members of
this tribe will become known to themselves and others as
the Wolves.

We need not rest satisfied with the inference that this
inheritance of nicknames *will* take place: there is proof
that it *does* take place. As nicknaming after animals,
plants, and other objects, still goes on among ourselves, so
among ourselves does there go on the descent of nicknames.
An instance has come under my own notice on an estate
in the West Highlands, belonging to some friends with
whom I frequently have the pleasure of spending a few
weeks in the autumn. "Take a young Croshck," has
more than once been the reply of my host to the inquiry,
who should go with me when I was setting out salmon-
fishing. The elder Croshck I knew well; and supposed
that this name, borne by him and by all belonging to him,
was the family surname. Some years passed before I
learned that the real surname was Cameron; that the

father was called Croshck, after the name of his cottage, to distinguish him from other Camerons employed about the premises; and that his children had come to be similarly distinguished. Though here, as very generally in Scotland, the nickname was derived from the place of residence, yet had it been derived from an animal, the process would have been the same—inheritance of it would have occurred just as naturally. Not even for this small link in the argument, however, need we depend on inference: there is fact to bear us out. Mr. Bates, in his "Naturalist on the River Amazon" (2d ed., p. 376), describing three half-castes who accompanied him on a hunting trip, says: "Two of them were brothers—namely, João (John) and Zephyrino Jabutí; Jabutí, or tortoise, being a nickname which their father had earned for his slow gait, and which, as is usual in this country, had descended as the surname of the family." Let me add the statement made by Mr. Wallace respecting this same region, that "one of the tribes on the river Isánna is called 'Jurupari' (Devils). Another is called 'Ducks;' a third, 'Stars;' a fourth, 'Mandiocca.'" Putting these two statements together, can there be any doubt about the genesis of these tribal names? Let the tortoise become sufficiently distinguished (not necessarily by superiority—great inferiority may occasionally suffice) and the tradition of descent from him, preserved by his descendants themselves if he was superior, and by their contemptuous neighbors if he was inferior, may become a tribal name.[1]

[1] Since the foregoing pages were written, my attention has been drawn by Sir John Lubbock to a passage in the appendix to the second edition of "Prehistoric Times," in which he has indicated this derivation of tribal names. He says: "In endeavoring to account for the worship of animals, we must remember that names are very frequently taken from them. The children and followers of a man called the Bear or the Lion would make that a tribal name. Hence the animal itself would be first respected, at last worshipped." Of the genesis of this worship, however, Sir John Lubbock does not give any specific

"But this," it will be said, "does not amount to an explanation of animal-worship." True: a third factor remains to be specified. Given a belief in the still-existing other self of the deceased ancestor, who must be propitiated; given this survival of his metaphorical name among his grandchildren, great-grandchildren, etc.; and the further requisite is that the distinction between metaphor and reality shall be forgotten. Let the tradition of the ancestor fail to keep clearly in view the fact that he was a man called the Wolf—let him be habitually spoken of as the Wolf, just as when alive; and the natural mistake of taking the name literally will bring with it, firstly, a belief in descent from the actual wolf, and, secondly, a treatment of the wolf in a manner likely to propitiate him—a manner appropriate to one who may be the other self of the dead ancestor, or one of the kindred, and therefore a friend.

That a misunderstanding of this kind will naturally grow up, becomes obvious when we bear in mind the great indefiniteness of primitive language. As Prof. Max Müller says, respecting certain misinterpretations of an opposite kind: "These metaphors . . . . would become mere names handed down in the conversation of a family, understood perhaps by the grandfather, familiar to the father, but strange to the son, and misunderstood by the grandson." We have ample reason, then, for thinking that such misinterpretations are likely to occur. Nay, we may go further. We are justified in saying that they are certain to occur. For undeveloped languages contain no words capable of indicating the distinction to be kept in view. In the tongues of existing inferior races, only con-

explanation. Apparently he inclines to the belief, tacitly adopted also by Mr. McLennan, that animal-worship is derived from an original Fetichism, of which it is a more developed form. As will shortly be seen, I take a different view of its origin.

crete objects and acts are expressible. The Australians have a name for each kind of tree, but no name for tree irrespective of kind. And though some witnesses allege that their vocabulary is not absolutely destitute of generic names, its extreme poverty in such is unquestionable. Similarly with the Tasmanians. Dr. Milligan says they "had acquired very limited powers of abstraction or generalization. They possessed no words representing abstract ideas; for each variety of gum-tree and wattle-tree, etc., etc., they had a name, but they had no equivalent for the expression, 'a tree;' neither could they express abstract qualities, such as hard, soft, warm, cold, long, short, round, etc.; for 'hard,' they would say 'like a stone,' for 'tall,' they would say 'long legs,' etc., and for 'round,' they said 'like a ball,' 'like the moon,' and so on, usually suiting the action to the word, and confirming, by some sign, the meaning to be understood."[1] Now, even making allowance for over-statement here (which seems needful, since the word "long," said to be inexpressible in the abstract, subsequently occurs as qualifying a concrete in the expression, "long legs"), it is sufficiently manifest that so imperfect a language must fail to convey the idea of a name, as something separate from a thing; and that still less can it be capable of indicating the act of naming. Familiar use of such partially abstract words as are applicable to all objects of a class, is needful before there can be reached the conception of a name—a word symbolizing the symbolic character of other words; and the conception of a name, with its answering abstract term, must be long current before the verb to name can arise. Hence, among tribes with speech so rude, it will be impossible to transmit the tradition of an ancestor named the Wolf, as distinguished from the actual wolf. The children and grandchildren who saw him will not be led into error; but in later generations, descent from the Wolf will inevitably come to mean descent from the animal known by that name. And the ideas and sentiments which, as above shown, naturally grow up around the belief that the dead parents and grandparents are still alive, and ready, if propitiated, to befriend their descendants, will be extended to the wolf species.

Before passing to other developments of this general view, let me point out how not simply animal-worship is thus accounted for, but also the conception, so variously illustrated in ancient legends, that animals are capable of displaying human powers of speech and thought and action. Mythologies are full of stories of beasts and birds and fishes that have played intelligent parts in human affairs—creatures that have befriended particular persons by giving them information, by guiding them, by yielding them help; or else that have deceived them, verbally or otherwise. Evidently all these traditions, as well as those about abductions of women by animals and fostering of children by them, fall naturally into their places as results of the habitual misinterpretation I have described.

The probability of the hypothesis will appear still greater when we observe how readily it applies to the worship of other orders of objects. Belief in actual descent from an animal, strange as we may think it, is one by no means incongruous with the unanalyzed experiences of the savage; for there come under his notice many metamorphoses, vegetal and animal, which are apparently of like character. But how could he possibly arrive at so grotesque a conception as that the progenitor of his tribe was the sun, or the moon, or a particular star? No observation of surrounding phenomena affords the slightest suggestion of any such possibility. But by the inheritance of nicknames that are eventually mistaken for the names of the objects from which they were derived, the belief readily arises—is sure to arise. That the names of heavenly bodies will furnish metaphorical names to the uncivilized, is manifest. Do we not ourselves call a distinguished singer or actor a star? And have we not in poems numerous comparisons of men and women to the sun and moon; as in "Love's Labour's Lost," where the princess is called "a gracious moon," and as in "Henry VIII.," where we read—"Those suns of glory, those two lights of men?" Clearly, primitive men will be not unlikely thus to speak of the chief hero of a successful battle. When we remember how the arrival of a triumphant warrior must affect the feelings of his tribe, dissipating clouds of anxiety and irradiating all faces with joy, we shall see that the comparison of him to the sun is extremely natural; and in early speech this comparison can be made only by calling him the sun. As before, then, it will happen that, through a confounding of the metaphorical name with the actual name, his progeny, after a few generations, will be regarded by themselves and others as descendants of the sun. And, as a consequence, partly of actual inheritance of the ancestral character, and partly of maintenance of the traditions respecting the ancestor's achievements, it will also naturally happen that the solar race will be considered a superior race, as we find it habitually is.

The origin of other totems, equally strange if not even stranger, is similarly accounted for, though otherwise unaccountable. One of the New-Zealand chiefs claimed as his progenitor the neighboring great mountain, Tongariro. This seemingly-whimsical belief becomes intelligible when we observe how easily it may have arisen from a nickname. Do we not ourselves sometimes speak figuratively of a tall, fat man as a mountain of flesh? And, among a people prone to speak in still more concrete terms, would it not happen that a chief, remarkable for his great bulk, would be nicknamed after the highest mountain within sight, because he towered above other men as this did above surrounding hills? Such an occurrence is not simply possible, but probable. And, if so, the confusion of metaphor with fact would originate this surprising genealogy. A notion perhaps yet more grotesque, thus receives a satisfactory interpretation. What could have put it into the imagination of any one that he was descended from the dawn? Given the extremest credulity, joined with the wildest fancy, it would still seem requisite that the ancestor should be conceived as an entity; and the dawn is entirely without that definiteness and comparative constancy which enter into the conception of an entity. But when we remember that "the Dawn" is a natural complimentary name for a beautiful girl opening into womanhood, the genesis of the idea becomes, on the above hypothesis, quite obvious.

Another indirect verification is that we thus get a clear conception of Fetichism in general. Under the fetichistic mode of thought, surrounding objects and agents are regarded as having powers more or less definitely personal in their natures. And the current interpretation is, that human intelligence, in its early stages, is obliged to conceive of their powers under this form. I have myself hitherto accepted this interpretation; though always with a sense of dissatisfaction. This dissatisfaction was, I think, well grounded. The theory is scarcely a theory properly so called; but rather, a restatement in other words. Uncivilized men *do* habitually form anthropomorphic conceptions of surrounding things; and this observed general fact is transformed into the theory that at first they *must* so conceive them—a theory for which the psychological justification attempted, seems to me inadequate. From our present stand-point, it becomes manifest that Fetichism is not primary but secondary. What has been said above almost of itself shows this. Let us, however, follow out the steps of its genesis. Respecting the Tasmanians, Dr. Milligan says: "The names of men and women were taken from natural objects and occurrences around, as, for instance, a kangaroo, a gum-tree, snow, hail, thunder, the wind, flowers in blossom, etc." Surrounding objects, then, giving origin to names of persons, and being, in the way shown, eventually mistaken for the actual progenitors of those who descend from persons nicknamed after them, it results that these surrounding objects come to be regarded as in some manner possessed of personalities like the human. He whose family tradition is that his ancestor was "the Crab," will conceive the crab as having a disguised inner power like his own; and alleged descent from "the palm-tree" will entail belief in some kind of consciousness dwelling in the palm-tree. Hence, in proportion as the animals, plants, and inanimate objects or agents that originate names of persons, become numerous (which they will do in proportion as a tribe becomes large and the number of persons to be distinguished from one another increases), multitudinous things around will acquire imaginary personalities. And so it will happen that, as Mr. McLennan says of the Feejeeans : "Vegetables and stones, nay, even tools and weapons, pots and canoes, have souls that are immortal, and that, like the souls of men, pass on at last to Mbulu, the abode of departed spirits." Setting out, then, with a belief in the still-living other self of the dead ancestor, the alleged general cause of misapprehension affords us an intelligible origin of the fetichistic conception; and we are enabled to see how it tends to become a general, if not a universal, conception.

Other apparently inexplicable phenomena are at the same time divested of their strangeness. I refer to the beliefs in, and worship of, compound monsters—impossible hybrid animals, and forms that are half human, half brutal. The theory of a primordial Fetichism, supposing it otherwise adequate, yields no feasible solution of these. Grant the alleged original tendency to think of all natural agencies as in some way personal. Grant, too, that hence may arise a worship of animals, plants, and even inanimate bodies. Still the obvious implication is that the worship so derived will be limited to things that are, or have been, perceived. Why should this mode of thought lead the savage to imagine a combination of bird and mammal; and not only to imagine it, but worship it as a god? If even we admit that some illusion may have suggested the belief in a creature half man, half fish, we cannot thus explain the prevalence among Eastern races of idols representing bird-headed men, men having their legs replaced by the legs of a cock, and men with the heads of elephants.

Carrying with us the inferences above drawn, however, it is a manifest corollary that ideas and practices of these kinds will arise. When tradition preserves both lines of ancestry—when a chief, nicknamed the Wolf, carries away from an adjacent tribe a wife who is remembered either under the animal name of her tribe, or as a woman; it will happen that if a son distinguishes himself, the remembrance of him among his descendants will be that he was born of a wolf and some other animal, or of a wolf and a woman. Misinterpretation, arising in the way described from defects of language, will entail belief in a creature uniting the attributes of the two; and if the tribe grows into a society, representations of such a creature will become objects of worship. One of the cases cited by Mr. McLennan may here be repeated in illustration. "The story of the origin of the Dikokamenni Kirghcez," they say, "from a red greyhound and a certain queen with her forty handmaidens, is of ancient date." Now, if "the red greyhound" was the nickname of a man extremely swift of foot (celebrated runners have been similarly nicknamed among ourselves), a story of this kind would naturally arise; and if the metaphorical name was mistaken for the actual name, there might result, as the idol of the race, a compound form appropriate to the story. We need not be surprised, then, at finding among the Egyptians the goddess Pasht represented as a woman with a lion's head, and the god Month as a man with the head of a hawk. The Babylonian gods—one having the form of a man with an eagle's tail, and another uniting a human bust to a fish's body—no longer appear such unaccountable conceptions. We get feasible explanations, too, of sculptures representing sphinxes, winged human-headed bulls, etc.; as well as of the stories about centaurs, satyrs, and the rest.

Ancient myths in general thus acquire meanings considerably different from those ascribed to them by comparative mythologists. Though these last may be in part correct, yet if the foregoing argument is valid, they can scarcely be correct in their main outlines. Indeed, if we read the facts the other way upward, regarding as secondary or additional the elements that are said to be primary, while we regard as primary certain elements which are considered as accretions of later times, we shall, I think, be nearer the truth.

The current theory of the myth is that it has grown out of the habit of symbolizing natural agents and processes, in terms of human personalities and actions. Now, it may in the first place be remarked that, though symbolization of this kind is common enough among civilized races, it is not common among races that are the most uncivilized. By existing savages, surrounding objects, motions, and changes, are habitually used to convey ideas respecting human transactions. It is by no means so much the habit to express by the doings of men the course of natural phenomena. It needs but to read the speech of an Indian chief to see that just as primitive men name one another metaphorically after surrounding objects, so do they metaphorically describe one another's doings as though they were the doings of natural objects. But assuming a contrary habit of thought to be the dominant one, ancient myths are explained as results of the primitive tendency to symbolize inanimate things and their changes, by human beings and their doings.

A kindred difficulty must be added. The change of verbal meaning from which the myth is said to arise, is a change opposite in kind to that which prevails in the earlier stages of linguistic development. It implies a derivation of the concrete from the abstract; whereas at first abstracts are derived only from concretes: the concreting of abstracts being a subsequent process. In the words of Prof. Max Müller, there are " dialects spoken at the present day which have no abstract nouns, and the more we go back in the history of languages, the smaller we find the number of these useful expressions " (" Chips," vol. ii., p. 54); or, as he says more recently: "Ancient words and ancient thoughts, for both go together, have not yet arrived at that stage of abstraction in which, for instance, active powers, whether natural or supernatural, can be represented in any but a personal and more or less human form." (*Fraser's Magazine*, April, 1870.) Here the concrete is represented as original, and the abstract as derivative. Immediately afterward, however, Prof. Max Müller, having given as examples of abstract nouns, "day and night, spring and winter, dawn and twilight, storm and thunder," goes on to argue that, "as long as people thought in language, it was simply impossible to speak of morning or evening, of spring and winter, without giving to these conceptions something of an individual, active, sexual, and at last personal character." ("Chips," etc., vol. ii., p. 55.) Here the concrete is derived from the abstract—the personal conception is represented as coming *after* the impersonal conception; and through such transformation of the impersonal into the personal, Prof. Max Müller considers ancient myths to have arisen. How are these propositions reconcilable? One of two things must be said: If originally there were none of these abstract nouns, then the earliest statements respecting the daily course of Nature were made in concrete terms—the personal elements of the myth were the primitive elements, and the impersonal expressions which are their equivalents came later. If this is not admitted, then it must be held that, until after there arose these abstract nouns, there were no current statements at all respecting these most conspicuous objects and changes which the heavens and the earth present; and that the abstract nouns having been somehow formed, and rightly formed, and used without personal meanings, afterward became personalized—a process the reverse of that which characterizes early linguistic progress.

No such contradictions occur if we interpret myths after the manner that has been indicated. Nay, besides escaping contradictions, we meet with unexpected solu-

3

tions.   The moment we try it, the key unlocks for us with
ease what seems a quite inexplicable fact, which the current hypothesis takes as one of its postulates.   Speaking
of such words as sky and earth, dew and rain, rivers and
mountains, as well as of the abstract nouns above named,
Prof. Max Müller says: "Now, in ancient languages every
one of these words had necessarily a termination expres-
sive of gender, and this naturally produced in the mind
the corresponding idea of sex, so that these names received
not only an individual but a sexual character.   There
was no substantive which was not either masculine or
feminine; neuters being of later growth, and distinguishable chiefly in the nominative." ("Chips," etc., vol. ii.,
p. 55.)   And this alleged necessity for a masculine or
feminine implication is assigned as a part of the reason
why these abstract nouns and collective nouns became
personalized.   But should not a true theory of these first
steps in the evolution of thought and language show us
how it happened that men acquired the seemingly-strange
habit of so framing their words for sky, earth, dew, rain,
etc., as to make them indicative of sex?   Or, at any rate,
must it not be admitted that an interpretation which, instead of assuming this habit to be "necessary," shows us
how it results, thereby acquires an additional claim to
acceptance?   The interpretation I have indicated does
this.   If men and women are habitually nicknamed, and
if defects of language lead their descendants to regard
themselves as descendants of the things from which the
names were taken, then masculine or feminine genders
will be ascribed to these things according as the ancestors
named after them were men or women.   If a beautiful
maiden known metaphorically as "the Dawn," afterward
becomes the mother of some distinguished chief called
"the North Wind," it will result that when, in course of

time, the two have been mistaken for the actual dawn and the actual north wind, these will, by implication, be respectively considered as male and female.

Looking, now, at the ancient myths in general, their seemingly most inexplicable trait is the habitual combination of alleged human ancestry and adventures, with the possession of personalities otherwise figuring in the heavens and on the earth, with totally non-human attributes. This enormous incongruity, not the exception but the rule, the current theory fails to explain. Suppose it to be granted that the great terrestrial and celestial objects and agents naturally become personalized; it does not follow that each of them shall have a specific human biography. To say of some star that he was the son of this king or that hero, was born in a particular place, and when grown up carried off the wife of a neighboring chief, is a gratuitous multiplication of incongruities already sufficiently great; and is not accounted for by the alleged necessary personalization of abstract and collective nouns. As looked at from our present stand-point, however, such traditions become quite natural—nay, it is clear that they will necessarily arise. When a nickname has become a tribal name, it thereby ceases to be individually distinctive; and, as already said, the process of nicknaming inevitably continues. It commences afresh with each child; and the nickname of each child is both an individual name and a potential tribal name, which may become an actual tribal name if the individual is sufficiently celebrated. Usually, then, there is a double system of distinguishing the individual; under one of which he is known by his ancestral name, and under the other of which he is known by a name suggestive of something peculiar to himself: just as we have seen happens among the Scotch clans. Consider, now, what will result when language

has reached a stage of development such that it can convey the notion of naming, and is able, therefore, to preserve traditions of human ancestry : the preservation of
such traditions being furthered by these corruptions of
tribal names which render them no longer suggestive of
the things they were derived from.   It will result that the
individual will be known both as the son of such and
such a man by a mother whose name was so-and-so, and
also as the Crab, or the Bear, or the Whirlwind—supposing one of these to be his nickname.   Such joint use of
nicknames and proper names occurs in every school.   Now,
clearly, in advancing from the early state in which ancestors become identified with the objects they are nicknamed after, to the state in which there are proper names
that have lost their metaphorical meanings, there must be
passed through a state in which proper names, partially
settled only, may or may not be preserved, and in which
the new nicknames are still liable to be mistaken for actual names.   Under such conditions there will arise (especially in the case of a distinguished man) this seeming-
ly-impossible combination of human parentage with the
possession of the non-human, or superhuman, attributes
of the thing which gave the nickname.   Another anomaly
simultaneously disappears.   The warrior may have, and
often will have, a variety of complimentary nicknames—
" the powerful one," " the destroyer," etc.   Supposing
his leading nickname has been the Sun, then when he
comes to be identified by tradition with the sun, it will
happen that the sun will acquire his alternative descriptive titles—the swift one, the lion, the wolf—titles not
obviously appropriate to the sun, but quite appropriate to
the warrior.   Then there comes, too, an explanation of
the remaining trait of such myths.   When this identification of conspicuous persons, male and female, with con-

spicuous natural agents, has become settled, there will in due course arise interpretations of the actions of these agents in anthropomorphic terms. Suppose, for instance, that Endymion and Selene, metaphorically named, the one after the setting sun, the other after the moon, have had their human individualities merged in those of the sun and moon, through misinterpretation of metaphors; what will happen? The legend of their loves having to be reconciled with their celestial appearances and motions, these will be spoken of as results of feeling and will; so that when the sun is going down in the west, while the moon in mid-heaven is following him, the fact will be expressed by saying: "Selene loves and watches Endymion." Thus we obtain a consistent explanation of the myth without distorting it; and without assuming that it contains gratuitous fictions. We are enabled to accept the biographical part of it, if not as literal fact, still as having had fact for its root. We are helped to see how, by an inevitable misinterpretation, there grew out of a more or less true tradition, this strange identification of its personages, with objects and powers totally non-human in their aspects. And then we are shown how, from the attempt to reconcile in thought these contradictory elements of the myth, there arose the habit of ascribing the actions of these non-human things to human motives.

One further verification may be drawn from facts which are obstacles to the converse hypothesis. These objects and powers, celestial and terrestrial, which force themselves most on men's attention, have some of them several proper names, identified with those of different individuals, born at different places, and having different sets of adventures. Thus we have the sun variously known as Apollo, Endymion, Helios, Tithonos, etc.—personages having irreconcilable genealogies. Such anoma-

lies Prof. Max Müller apparently ascribes to the untrust-
worthiness of traditions, which are " careless about contradictions, or ready to solve them sometimes by the most
atrocious expedients." (" Chips," etc., vol. ii., p. 84.) But
if the evolution of the myth has been that above indicated,
there exist no anomalies to be got rid of: these diverse
genealogies become parts of the evidence. For we have
abundant proof that the same objects furnish metaphorical names of men in different tribes. There are Duck
tribes in Australia, in South America, in North America.
The eagle is still a totem among the North Americans, as
Mr. McLennan shows reason to conclude that it was
among the Egyptians, among the Jews, and among the
Romans. Obviously, for reasons that have been assigned,
it naturally happened in the early stages of the ancient
races, that complimentary comparisons of their heroes to
the sun were frequently made. What resulted? The
sun having furnished names for sundry chiefs and early
founders of tribes, and local traditions having severally
identified them with the sun, these tribes, when they grew,
spread, conquered, or came otherwise into partial union,
originated a combined mythology, which necessarily contained conflicting stories about the sun-god, as about its
other leading personages. If the North-American tribes,
among several of which there are traditions of a sun-god,
had developed a combined civilization, there would similarly have arisen among them a mythology which ascribed
to the sun several different proper names and genealogies.

Let me briefly set down the leading characters of this
hypothesis which give it probability.

True interpretations of all the natural processes, organic and inorganic, that have gone on in past times,
habitually trace them to causes still in action. It is thus

in Geology ; it is thus in Biology ; it is thus in Philology. Here we find this characteristic repeated. Nicknaming, the inheritance of nicknames, and, to some extent, the misinterpretation of nicknames, go among us still ; and were surnames absent, language imperfect, and knowledge as rudimentary as of old, it is tolerably manifest that results would arise like those we have contemplated.

A further characteristic of a true cause is that it accounts not only for the particular group of phenomena to be interpreted, but also for other groups. The cause here alleged does this. It equally well explains the worship of animals, of plants, of mountains, of winds, of celestial bodies, and even of appearances too vague to be considered entities. It gives us an intelligible genesis of fetichistic conceptions in general. It furnishes us with a reason for the practice, otherwise so unaccountable, of moulding the words applied to inanimate objects in such ways as to imply masculine and feminine genders. It shows us how there naturally arose the worship of compound animals, and of monsters half man half brute. And it shows us why the worship of purely anthropomorphic deities came later, when language had so far developed that it could preserve in tradition the distinction between proper names and nicknames.

A further verification of this view is, that it conforms to the general law of evolution : showing us how, out of one simple, vague, aboriginal form of belief, there have arisen, by continuous differentiations, the many heterogeneous forms of belief which have existed and do exist. The desire to propitiate the other self of the dead ancestor, displayed among savage tribes, dominantly manifested by the early historic races, by the Peruvians and Mexicans, by the Chinese at the present time, and to a considerable degree by ourselves (for what else is the wish to do that which a lately-deceased parent was known to have desired?), has been the universal first form of religious belief; and from it have grown up the many divergent beliefs that have been referred to.

Let me add, as a further reason for adopting this view, that it immensely diminishes the apparently-great contrast between early modes of thought and our own mode of thought. Doubtless the aboriginal man differs considerably from us, both in intellect and feeling. But such an interpretation of the facts as helps us to bridge over the gap, derives additional likelihood from doing this. The hypothesis I have sketched out enables us to see that primitive ideas are not so gratuitously absurd as we suppose, and also enables us to rehabilitate the ancient myth with far less distortion than at first sight appears possible.

These views I hope to develop in the first part of " The Principles of Sociology." The large mass of evidence which I shall be able to give in support of the hypothesis, joined with the solutions it will be shown to yield of many minor problems which I have passed over, will, I think, then give to it a still greater probability than it seems now to have.

III.

THE CLASSIFICATION OF THE SCIENCES.

PREFACE TO THE SECOND EDITION.

THE first edition of this Essay is not yet out of print. But a proposal to translate it into French having been made by Professor Réthoré, I have decided to prepare a new edition free from the imperfections which criticism and further thought have disclosed, rather than allow these imperfections to be reproduced.

The occasion has almost tempted me into some amplification. Further arguments against the classification of M. Comte, and further arguments in support of the classification here set forth, have pleaded for utterance. But reconsideration has convinced me that it is both needless and useless to say more —needless because those who are not committed will think the case sufficiently strong as it stands, and useless because to those who are committed additional reasons will seem as inadequate as the original ones.

This last conclusion is thrust on me by seeing how little M. Littré, the leading expositor of M. Comte, is influenced by fundamental objections the force of which he admits. After quoting one of these, he

says, with a candour equally rare and admirable, that
he has vainly searched M. Comte's works and his
own mind for an answer. Nevertheless, he adds—
"j'ai réussi, je crois, à écarter l'attaque de M. Herbert Spencer, et à sauver le fond par des sacrifices
indispensables mais accessoires." The sacrifices are
these.   He abandons M. Comte's division of Inorganic Science into Celestial Physics and Terrestrial Physics—a division which, in M. Comte's
scheme, takes precedence of all the rest; and he
admits that neither logically nor historically does
Astronomy come before Physics, as M. Comte alleges.
After making these sacrifices, which most will think
too lightly described as "sacrifices indispensables
mais accessoires," M. Littré proceeds to rehabilitate
the Comtean classification in a way which he considers satisfactory, but which I do not understand.
In short, the proof of these incongruities affects his
faith in the Positivist theory of the sciences, no
more than the faith of a Christian is affected by
proof that the Gospels contradict one another.

Here in England I have seen no attempt to meet
the criticisms with which M. Littré thus deals.
There has been no reply to the allegation, based on
examples, that the several sciences do not develop
in the order of their decreasing generality; nor to
the allegation, based on M. Comte's own admissions,
that within each science the progress is not, as he
says it is, from the general to the special; nor to

the allegation that the seeming historical precedence of Astronomy over Physics in M. Comte's pages, is based on a verbal ambiguity—a mere sleight of words; nor to the allegation, abundantly illustrated, that a progression in an order the reverse of that asserted by M. Comte may be as well substantiated; nor to various minor allegations equally irreconcileable with his scheme. I have met with nothing more than iteration of the statement that the sciences *do* conform, logically and historically, to the order in which M. Comte places them; regardless of the assigned evidence that they *do not*.

Under these circumstances it is unnecessary for me to say more; and I think I am warranted in continuing to hold that the Comtean classification of the sciences is demonstrably untenable.

While, however, I have not entered further into the controversy, as I thought of doing, I have added at the close an already-published discussion, no longer easily accessible, which indirectly enforces the general argument.

LONDON, 23RD APRIL, 1869.

PREFACE TO THE THIRD EDITION.

In the preface to the second edition, I have described myself as resisting the temptation to amplify, which the occasion raised. Reasons have since arisen for yielding to the desire which I then felt to add justifications of the scheme set forth.

The immediate cause for this change of resolve, has been the publication of several objections by Prof. Bain in his Logic. Permanently embodied, as these objections are, in a work intended for the use of students, they demand more attention than such as have been made in the course of ordinary criticism; since, if they remain unanswered, their prejudicial effects will be more continuous.

While to dispose of these I seize the opportunity afforded by a break in my ordinary work, I have thought it well at the same time to strengthen my own argument, by a re-statement from a changed point of view.

Feb., 1871.

CLASSIFICATION OF THE SCIENCES.

IN an essay on " The Genesis of Science," originally published in 1854,[*] I endeavoured to show that the Sciences cannot be rationally arranged in serial order. Proof was given that neither the succession in which the Sciences are placed by M. Comte (to a criticism of whose scheme the essay was in part devoted), nor any other succession in which the Sciences can be placed, represents either their logical dependence or their historical dependence. To the question—How may their relations be rightly expressed? I did not then attempt any answer. This question I propose now to consider.

A true classification includes in each class, those objects which have more characteristics in common with one another, than any of them have in common with any objects excluded from the class. Further, the characteristics possessed in common by the colligated objects, and not possessed by other objects, are more radical than any characteristics possessed in common with other objects—involve more numerous

---

[*] Contained in the " Illustrations of Universal Progress."

dependent characteristics. These are two sides of the
same definition. For things possessing the greatest
number of attributes in common, are things that possess in common those essential attributes on which the
rest depend ; and, conversely, the possession in common of the essential attributes, implies the possession
in common of the greatest number of attributes. Hence,
either test may be used as convenience dictates.

   If, then, the Sciences admit of classification at all, it
must be by grouping together the like and separating
the unlike, as thus defined. Let us proceed to do this.

   The broadest natural division among the Sciences,
is the division between those which deal with the abstract relations under which phenomena are presented
to us, and those which deal with the phenomena themselves. Relations of whatever orders, are nearer akin
to one another than they are to any objects. Objects
of whatever orders, are nearer akin to one another
than they are to any relations. Whether, as some
hold, Space and Time are forms of Thought; or
whether, as I hold myself, they are forms of Things,
that have become forms of Thought through organized and inherited experience of Things; it is equally
true that Space and Time are contrasted absolutely
with the existences disclosed to us in Space and Time
and that the Sciences which deal exclusively with
Space and Time, are separated by the profoundest of
all distinctions from the Sciences which deal with the

existences that Space and Time contain. Space is the abstract of all relations of co-existence. Time is the abstract of all relations of sequence. And dealing as they do entirely with relations of co-existence and sequence, in their general or special forms, Logic and Mathematics form a class of the Sciences more widely unlike the rest, than any of the rest can be from one another.

The Sciences which deal with existences themselves, instead of the blank forms in which existences are presented to us, admit of a sub-division less profound than the division above made, but more profound than any of the divisions among the Sciences individually considered. They fall into two classes, having quite different aspects, aims, and methods. Every phenomenon is more or less composite—is a manifestation of force under several distinct modes. Hence result two objects of inquiry. We may study the component modes of force separately; or we may study them in their relations, as co-operative factors in this composite phenomenon. On the one hand, neglecting all the incidents of particular cases, we may aim to educe the laws of each mode of force, when it is uninterfered with. On the other hand, the incidents of the particular case being given, we may seek to interpret the entire phenomenon, as a product of the several forces simultaneously in action. The truths reached through the first kind of inquiry, though concrete inasmuch as they have actual existences for their subject-matters, are abstract inasmuch as they refer to the modes of existence apart from one another; while the truths reached by the second kind of inquiry are properly concrete, inasmuch as they formulate the facts in their combined order, as they occur in Nature.

The Sciences, then, in their main divisions, stand thus :—

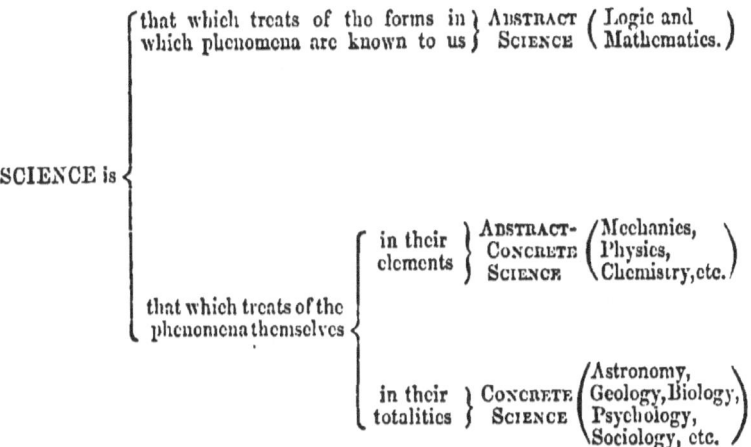

It is needful to define the words *abstract* and *concrete* as thus used ; since they are sometimes used with other meanings. M. Comte divides Science into abstract and concrete; but the divisions which he distinguishes by these names are quite unlike those above made. Instead of regarding some Sciences as wholly abstract, and others as wholly concrete, he regards each Science as having an abstract part, and a concrete part. There is, according to him, an abstract mathematics and a concrete mathematics—an

abstract biology and concrete biology. He says:—
"Il faut distinguer, par rapport à tous les ordres de
phénomènes, deux genres de sciences naturelles: les
unes abstraites, générales, ont pour objet la découverte
des lois qui régissent les diverses classes de phénomènes, en considérant tous les cas qu'on peut con-
cevior ; les autres concrètes, particulières, descriptives,
et qu'on désigne quelquefois sous le nom de sciences
naturelles proprement dites, consistent dans l'application de ces lois a l'histoire effective de différens êtres
existans." And to illustrate the distinction, he names
general physiology as abstract, and zoology and botany
as concrete. Here it is manifest that the words
abstract and *general* are used as synonymous. They
have, however, different meanings ; and confusion
results from not distinguishing their meanings. Abstractness means *detachment from* the incidents of particular cases. Generality means *manifestation in* numerous
cases. On the one hand, the essential nature of some
phenomenon is considered, apart from disguising phenomena. On the other hand, the frequency of the
phenomenon, with or without disguising phenomena,
is the thing considered. Among the ideal relations of
numbers the two coincide ; but excluding these, an
abstract truth is not realizable to perception in any
case in which it is asserted, whereas a general truth is
realizable to perception in every case of which it is
asserted. Some illustrations will make the distinction
clear. Thus it is an abstract truth that the angle contained

in a semi-circle is a right angle—abstract in the sense
that though it does not hold in actually-constructed
semi-circles and angles, which are always inexact, it
holds in the ideal semi-circles and angles abstracted
from real ones; but this is not a general truth, either
in the sense that it is commonly manifested in Nature,
or in the sense that it is a space-relation that compre-
hends many minor space-relations : it is a quite
special space-relation. Again, that the momentum
of a body causes it to move in a straight line at a
uniform velocity, is an abstract-concrete truth—a
truth abstracted from certain experiences of concrete
phenomena; but it is by no means a general truth :
so little generality has it, that no one fact in Nature
displays it. Conversely, surrounding things supply
us with hosts of general truths that are not in the
least abstract. It is a general truth that the planets
go round the Sun from West to East—a truth which
holds good in something like a hundred cases (including the cases of the planetoids); but this truth
is not at all abstract, since it is perfectly realized
as a concrete fact in every one of these cases. Every
vertebrate animal whatever, has a double nervous
system ; all birds and all mammals are warm-
blooded—these are general truths, but they are
concrete truths : that is to say, every vertebrate
animal individually presents an entire and unqualified
manifestation of this duality of the nervous system;
every living bird exemplifies absolutely or completely

the warm-bloodedness of birds. What we here call, and rightly call, a general truth, is simply a proposition which *sums up* a number of our actual experiences ; and not the expression of a truth *drawn from* our actual experiences, but never presented to us in any of them. In other words, a general truth colligates a number of particular truths ; while an abstract truth colligates no particular truths, but formulates a truth which certain phenomena all involve, though it is actually seen in none of them.

Limiting the words to their proper meanings as thus defined, it becomes manifest that the three classes of Sciences above separated, are not distinguishable at all by differences in their degrees of generality. They are all equally general ; or rather they are all, considered as groups, universal. Every object whatever presents at once the subject-matter for each of them. In the smallest particle of substance we have simultaneously illustrated the abstract truths of relation in Time and Space ; the abstract-concrete truths in conformity with which the particle manifests its several modes of force ; and the concrete truths which are the laws of the joint manifestation of these modes of force. Thus these three classes of Sciences severally formulate different, but co-extensive, classes of facts. Within each group there are truths of greater and less generality : there are general abstract truths, and special abstract truths ; general abstract-concrete truths, and special abstract-concrete truths ·

general concrete truths, and special concrete truths.
But while within each class there are groups and
sub-groups and sub-sub-groups which differ in their
degrees of generality, the classes themselves differ
only in their degrees of abstractness.*

---

* Some propositions laid down by M. Littré, in his lately-published book—
Auguste Comte et la Philosophie Positive, may fitly be dealt with here. In the
candid and courteous reply he makes to my strictures on the Comtean classification in "The Genesis of Science," he endeavours to clear up some of the incon-
sistencies I pointed out; and he does this by drawing a distinction between
objective generality and subjective generality.   He says—"qu'il existe deux
ordres de généralité, l'une objective et dans les choses, l'autre subjective, abstraite
et dans l'esprit." This sentence, in which M. Littré makes subjective generality
synonymous with abstractness, led me at first to conclude that he had in view the
same distinction as that which I have above explained between generality and
abstractness.   On re-reading the paragraph, however, I found this was not the
case.   In a previous sentence he says—"La biologie a passé de la considération
des organes à celles des tissus, plus généraux que les organes, et de la considération
des tissus à celle des éléments anatomiques, plus généraux que les tissus. Mais
cette généralité croissante est subjective non objective, abstraite non concrète."
Here it is manifest that abstract and concrete, are used in senses analogous to
those in which they are used by M. Comte; who, as we have seen, regards
general physiology as abstract and zoology and botany as concrete. And it is
further manifest that the word abstract, as thus used, is not used in its proper
sense. For, as above shown, no such facts as those of anatomical structure can
be abstract facts; but can only be more or less general facts   Nor do I understand M. Littré's point of view when he regards these more general facts of
anatomical structure, as *subjectively* general and not *objectively* general.   The
structural phenomena presented by any tissue, such as mucous membrane, are
more general than the phenomena presented by any of the organs which mucous
membrane goes to form, simply in the sense that the phenomena peculiar to the
membrane are repeated in a greater number of instances than the phenomena
peculiar to any organ into the composition of which the membrane enters.   And,
similarly, such facts as have been established respecting the anatomical elements
of tissues, are more general than the facts established respecting any particular
tissue, in the sense that they are facts which organic bodies exhibit in a greater
number of cases—they are *objectively* more general; and they can be called
subjectively more general only in the sense that the conception corresponds with
the phenomena.

Let me endeavour to clear up this point:—There is, as M. Littré truly says,
a decreasing generality that is objective. If we omit the phenomena of Dissolution, which are changes from the special to the general, all changes which matter
undergoes are from the general to the special—are changes involving a decreasing

Passing to the sub-divisions of these classes, we find that the first class is separable into two parts—the one containing universal truths, the other non-universal truths. Dealing wholly with relations apart from related things, Abstract Science considers first, that which is common to all relations whatever; and second, that which is common to each order of relations. Besides the indefinite and variable connexions which exist among phenomena, as occurring together in Space and Time, we find that there are also definite generality in the united groups of attributes. This is the progress of *things*. The progress of *thought*, is not only in the same direction, but also in the opposite direction. The investigation of Nature discloses an increasing number of specialities; but it simultaneously discloses more and more the generalities within which these specialities fall. Take a case. Zoology, while it goes on multiplying the number of its species, and getting a more complete knowledge of each species (decreasing generality); also goes on discovering the common characters by which species are united into larger groups (increasing generality). Both these are subjective processes; and in this case, both orders of truths reached are concrete—formulate the phenomena as actually manifested.

M. Littré, recognizing the necessity for some modification of the hierarchy of the Sciences, as enunciated by M. Comte, still regards it as substantially true; and for proof of its validity, he appeals mainly to the essential *constitutions* of the Sciences. It is unnecessary for me here to meet, in detail, the arguments by which he supports the proposition, that the essential constitutions of the Sciences, justify the order in which M. Comte places them. It will suffice to refer to the foregoing pages, and to the pages which are to follow, as containing the definitions of those fundamental characteristics which demand the grouping of the Sciences in the way pointed out. As already shown, and as will be shown still more clearly by and bye, the radical differences of constitution among the Sciences, necessitate the colligation of them into the three classes—Abstract, Abstract-Concrete, and Concrete. How irreconcilable is M. Comte's classification with these groups, will be at once apparent on inspection. It stands thus:—

Mathematics (including rational Mechanics), ... partly Abstract, partly Abstract-Concrete.
Astronomy .... Concrete.
Physics.. Abstract-Concrete.
Chemistry .... Abstract-Concrete.
Biology... Concrete.
Sociology .... Concrete.

and invariable connexions—that between each kind of phenomenon and certain other kinds of phenomena, there exist uniform relations.   This is a universal abstract truth—that there is an unchanging order among things in Space and Time. We come next to the several kinds of unchanging order, which, taken together, form the subjects of the second division of Abstract Science.   Of this second division, the most general sub-division is that which deals with the natures of the connexions in Space and Time, irrespective of the terms connected.   The conditions under which we may predicate a relation of coincidence or proximity in Space and Time (or of non-coincidence or non-proximity) form the subject-matter of Logic.   Here the natures and amounts of the terms between which the relations are asserted (or denied) are of no moment : the propositions of Logic are independent of any qualitative or quantitative specification of the related things. The other sub-division has for its subject-matter, the relations between terms which are specified quantitatively but not qualitatively.   The amounts of the related terms, irrespective of their natures, are here dealt with; and Mathematics is a statement of the laws of quantity considered apart from reality.   Quantity considered apart from reality, is occupancy of Space or Time; and occupancy of Space or Time is measured by the number of coexistent or sequent positions occupied.   That is to say, quantities can be compared and the relations between them established, only by some direct or indirect enumeration of their component units; and the ultimate units into which all others are decomposable, are such occupied positions in Space as can, by making impressions on consciousness, produce occupied positions in Time. Among units that are unspecified in their natures (extensive, protensive, or intensive), but are ideally endowed with existence considered apart from attributes, the quantitative relations that arise, are those most general relations expressed by numbers. Such relations fall into either of two orders, according as the units are considered simply as capable of filling separate places in consciousness, or according as they are considered as filling places that are not only separate, but equal. In the one case, we have that indefinite calculus by which numbers of abstract existences, but not sums of abstract existence, are predicable. In the other case, we have that definite calculus by which both numbers of abstract existences and sums of abstract existence are predicable. Next comes that division of Mathematics which deals with the quantitative relations of magnitudes (or aggregates of units) considered as coexistent, or as occupying Space—the division called Geometry. And then we arrive at relations, the terms of which include both quantities of Time and quantities of Space—those in which times are estimated by the units of space traversed at a uniform velocity, and those in which equal

4

units of time being given, the spaces traversed with uniform or variable velocities are estimated. These Abstract Sciences, which are concerned exclusively with relations and with the relations of relations, may be grouped as shown in Table I.

Passing from the Sciences that treat of the ideal or unoccupied forms of relations, and turning to the Sciences that treat of real relations, or the relations among realities, we come first to those Sciences which deal with realities, not as they are habitually manifested to us, but with realities as manifested in their different modes, when these are artificially separated from one another. In the same way that the Abstract Sciences are ideal, relatively to the Abstract-Concrete and Concrete Sciences; so the Abstract-Concrete Sciences are ideal, relatively to the Concrete Sciences. Just as Logic and Mathematics have for their object to generalize the laws of relation, qualitative and quantitative, apart from related things; so, Mechanics, Physics, Chemistry, etc., have for their object to generalize the laws of relation which different modes of Matter and Motion conform to, when severally disentangled from those actual phenomena in which they are mutually modified. Just as the geometrician formulates the properties of lines and surfaces, independently of the irregularities and thicknesses of lines and surfaces as they really exist; so, the physicist and the chemist formulate the mani-

odes of Being, irrespective of any specification of the

or proximity in Time and Space, but not necessarily in

sets of positions in space; and the facts predicated being

independent existences.

when their numbers are completely specified
(*Arithmetic.*)

when their numbers
are specified only
— in their relations.
(*Algebra.*)

— in the relations of their relations.
(*Calculus of Operations.*)

nsidered in their relations of coexistence.
(*Geometry.*)

nsidered as traversed in Time
— that is wholly indefinite.
(*Kinematics.*)

— that is divided into equal units.
(*Geometry of Motion.*‡)

ABSTRACT SCIENCE.

TABLE I.

Universal law of relation—an expression of the truth that uniformities of c‹
natures of the uniformities of connexion.

Laws of relations

that are qualitative; or that are specified in their nature‹
their terms: the natures and amount of which are indiffe‹

that are quantitative
(MATHEMATICS)

negatively: the terms of the relat‹
the absences of certain quantities.

positively: the
terms being magni-
tudes composed of

units that ‹

equal units

* This definition includes the laws of relations called necessary, but not those of
relations called contingent. These last, in
which the probability of an inferred connexion varies with the number of times such
c nnexion has occurred in experience, are
rightly dealt with mathematically.

** Here, by way of explanation of the term negatively-quantitative, it
will suffice to instance the proposition that certain three lines will meet
in a point, as a negatively-quantitative proposition; since it asserts the
absence of any quantity of space between their intersections. Similarly,
the assertion that certain three points will always fall in a straight
line, is negatively-quantitative; since the conception of a straight line
implies the negation of any lateral quantity, or deviation.

† Lest the meaning of this division should not be understood, it may be well to
name, in illustration, the estimates of the statistician. Calculations respecting population, crime, disease, etc., have results which are correct only numerically, and not
in respect of the totalities of being or action represented by the numbers.

‡ Perhaps it will be asked—How can there be a Geometry of Motion into which the conception of Force does not enter? The reply is, that the time-relations and space-relations of
Motion may be considered apart from those of Force, in the same way that the space-relation‹
of Matter may be considered apart from Matter.

xion obtain among modes of Being, irrespective of any specification of the

relations of coincidence or proximity in Time and Space, but not necessarily in
. (LOGIC.) *

being definitely-related sets of positions in space; and the facts predicated being
ometry *of Position.* * *)

equal only as having independent existences.
(*Indefinite Calculus.*†)

ie equality of which is
ot defined as extensive,
rotensive, or intensive
(*Definite Calculus*) { when their numbers are completely specified
(*Arithmetic.*)

when their numbers
are specified only { in their relations.
(*Algebra.*)

in the relations of their relations.
(*Calculus of Operations.*)

he equality of which
s that of extension { considered in their relations of coexistence.
(*Geometry.*)

considered as traversed in Time { that is wholly indefinite.
(*Kinematics.*)

that is divided into equal
units.
(*Geometry of Motion.*‡)

festations of each mode of force, independently of the disturbances in its manifestations which other modes of force cause in every actual case. In works on Mechanics, the laws of motion are expressed without reference to friction and resistance of the medium. Not what motion ever really is, but what it would be if retarding forces were absent, is asserted. If any retarding force is taken into account, then the effect of this retarding force is alone contemplated: neglecting the other retarding forces. Consider, again, the generalizations of the physicist respecting molecular motion. The law that light varies inversely as the square of the distance, is absolutely true only when the radiation goes on from a point without dimensions, which it never does; and it also assumes that the rays are perfectly straight, which they cannot be unless the medium differs from all actual media in being perfectly homogeneous. If the disturbing effects of changes of media are investigated, the formulæ expressing the refractions take for granted that the new media entered are homogeneous; which they never really are. Even when a compound disturbance is allowed for, as when the refraction undergone by light in traversing a medium of increasing density, like the atmosphere, is calculated, the calculation still supposes conditions that are unnaturally simple—it supposes that the atmosphere is not pervaded by heterogeneous currents, which it always is. Similarly with the inquiries of the chemist. He does not take his substances as Nature supplies them. Before he proceeds to specify their respective properties, he purifies them—separates from each all trace of every other. Before ascertaining the specific gravity of a gas, he has to free this gas from the vapour of water, usually mixed with it. Before describing the properties of a salt, he guards against any error that may arise from the presence of an uncombined portion of the acid or base. And when he alleges of any element that it has a certain atomic weight, and unites with such and such equivalents of other elements, he does not mean that the results thus expressed are exactly the results of any one experiment; but that they are the results which, after averaging many trials, he concludes would be realized if absolute purity could be obtained, and if the experiments could be conducted without loss. His problem is to ascertain the laws of combination of molecules, not as they are actually displayed, but as they would be displayed in the absence of those minute interferences which cannot be altogether avoided. Thus all these Abstract-Concrete Sciences have for their object, *analytical interpretation.* In every case it is the aim to decompose the phenomenon, and formulate its components apart from one another; or some two or three apart from the rest. Wherever, throughout these Sciences, synthesis is employed, it is for the verification of analysis.*

* I am indebted to Prof. Frankland for reminding me of an objection that may be

The truths elaborated are severally asserted, not as truths exhibited by this or that particular object; but as truths universally holding of Matter and Motion in their more general or more special forms, considered apart from particular objects, and particular places in space.

The sub-divisions of this group of Sciences, may be drawn on the same principle as that on which the sub-divisions of the preceding group were drawn. Phenomena, considered as more or less involved manifestations of force, yield on analysis, certain laws of manifestation that are universal, and other laws of manifestation, which, being dependent on conditions, are not universal. Hence the Abstract-Concrete Sciences are primarily divisible into—the laws of force considered apart from its separate modes, and laws of force considered under each of its separate modes. And this second division of the Abstract-Concrete group, is sub-divisible after a manner essentially analogous. It is needless to occupy space by

made to this statement. The production of new compounds by synthesis, has of late become an important branch of chemistry. According to certain known laws of composition, complex substances, which never before existed, are formed, and fulfil anticipations both as to their general properties and as to the proportions of their constituents—as proved by analysis. Here it may be said with truth, that analysis is used to verify synthesis. Nevertheless, the exception to the above statement is apparent only—not real. In so far as the production of new compounds is carried on merely for the obtainment of such new compounds, it is not Science but Art—the application of pre-established knowledge to the achievement of ends. The proceeding is a part of Science, only in so far as it is a means to the better interpretation of the order of Nature. And how does it aid the interpretation? It does it only by verifying the pre-established conclusions respecting the laws of molecular combination; or by serving further to explain them. That is to say, these syntheses, considered on their scientific side, have simply the purpose of *forwarding the analysis of the laws of chemical combination.*

defining these several orders and genera of Sciences.
Table II. will sufficiently explain their relations.

We come now to the third great group. We have
done with the Sciences which are concerned only with
the blank forms of relations under which Being is
manifested to us. We have left behind the Sciences
which, dealing with Being under its universal mode,
and its several non-universal modes regarded as independent, treats the terms of its relations as simple and
homogeneous, which they never are in Nature. There
remain the Sciences which, taking these modes of
Being as they are connected with one another, have for
the terms of their relations, those heterogeneous combinations of forces that constitute actual phenomena.
The subject-matter of these Concrete-Sciences is the
real, as contrasted with the wholly or partially ideal.
It is their aim, not to separate and generalize apart
the components of all phenomena; but to explain each
phenomenon as a product of these components. Their
relations are not, like those of the simplest Abstract-
Concrete Sciences, relations between one antecedent
and one consequent, nor are they, like those of the
more involved Abstract-Concrete Sciences, relations
between some few antecedents cut off in imagination
from all others, and some few consequents similarly
cut off; but they are relations each of which has for
its terms a complete plexus of antecedents and a complete plexus of consequents. This is manifest in the

ns of resolution and composition of forces.

lid.   (*Statics.*)

iid.   (*Hydrostatics.*)

lid.   (*Dynamics.*)

iid.   (*Hydrodynamics.*)

eneral, as impenetrability or space-occupancy.

ecial, as the forms resulting from molecular equilibrium.

tter (cohesion, elasticity, etc.) $\left\{\begin{array}{l} \text{when solid.} \\ \text{when liquid.} \\ \text{when gaseous.} \end{array}\right.$

their relative omogeneously $\left\{\begin{array}{l} \text{causing increase of volume} \\ \text{(expansion, liquefaction, evaporation).} \\ \\ \text{causing decrease of volume} \\ \text{(condensation, solidification, contraction).} \end{array}\right.$

their relative terogeneously iistry) $\left\{\begin{array}{l} \text{producing new relations of molecules} \\ \text{(new compounds).} \\ \\ \text{producing new relations of forces} \\ \text{(new affinities).} \end{array}\right.$

ich, by integration, generates sensible motion.

ich, by disintegration, generates nsible motion, under the forms of $\left\{\begin{array}{l} \textit{Heat.} \\ \textit{Light.} \\ \textit{Electricity.} \\ \textit{Magnetism.} \end{array}\right.$

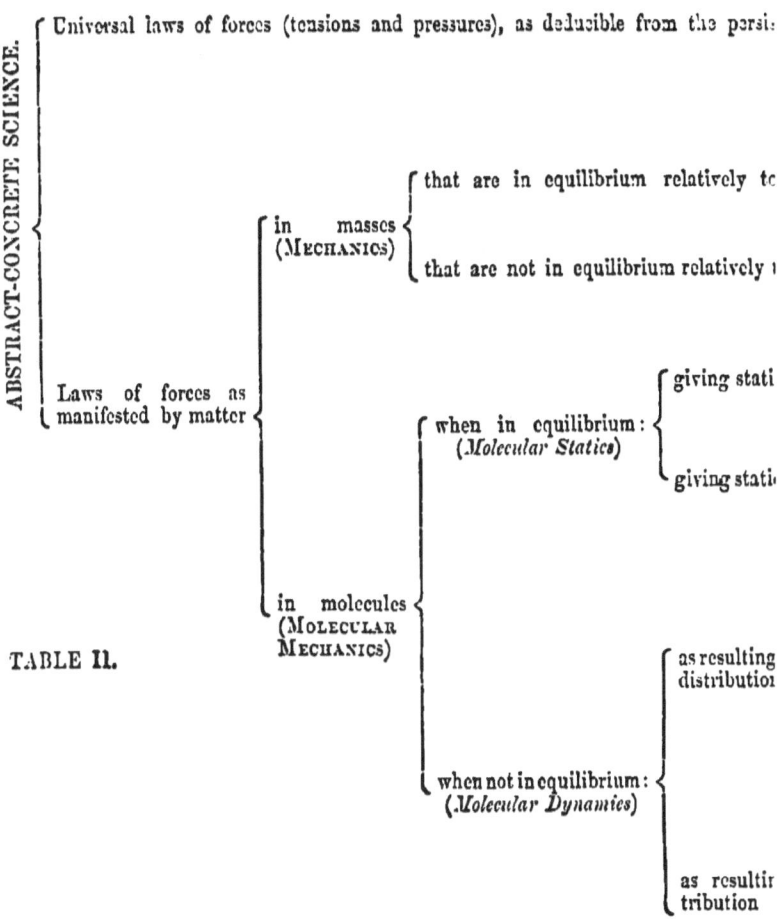

ABSTRACT-CONCRETE SCIENCE.

{ Universal laws of forces (tensions and pressures), as deducible from the persi⸱

Laws of forces as manifested by matter {

in masses (MECHANICS) {

that are in equilibrium relatively to

that are not in equilibrium relatively t

in molecules (MOLECULAR MECHANICS) {

when in equilibrium: (Molecular Statics) {

giving stati

giving stati⸱

when not in equilibrium: (Molecular Dynamics) {

as resulting distribution

as resultir tribution

TABLE II.

73

e of force: the theorems of resolution and composition of forces.

her masses { and are solid. (*Statics.*)
{ and are fluid. (*Hydrostatics.*)

her masses { and are solid. (*Dynamics.*)
{ and are fluid. (*Hydrodynamics.*)

properties of matter { general, as impenetrability or space-occupancy.
{ special, as the forms resulting from molecular equilibrium.

ynamical properties of matter (cohesion, elasticity, etc.) { when solid.
{ when liquid.
{ when gaseous.

changed
molecules
{ which alters their relative positions homogeneously { causing increase of volume (expansion, liquefaction, evaporation).
{ causing decrease of volume (condensation, solidification, contraction).

{ which alters their relative positions heterogeneously (*Chemistry*) { producing new relations of molecules (new compounds).
{ producing new relations of forces (new affinities).

n a changed dis-
molecular motion. { which, by integration, generates sensible motion.

{ which, by disintegration, generates insensible motion, under the forms of { *Heat.*
{ *Light.*
{ *Electricity.*
{ *Magnetism.*

least involved Concrete Sciences. The astronomer seeks to explain the Solar System. He does not stop short after generalizing the laws of planetary movement, such as planetary movement would be did only a single planet exist; but he solves this abstract-concrete problem, as a step towards solving the concrete problem of the planetary movements as affecting one another. In astronomical language, "the theory of the Moon" means an interpretation of the Moon's motions, not as determined simply by centripetal and centrifugal forces, but as perpetually modified by gravitation towards the Earth's equatorial protuberance, towards the Sun, and even towards Venus—forces daily varying in their amounts and combinations. Nor does the astronomer leave off when he has calculated what will be the position of a given body at a given time, allowing for all perturbing influences; but he goes on to consider the effects produced by reactions on the perturbing masses. And he further goes on to consider how these mutual perturbations of the planets cause, during a long period, increasing deviations from a mean state; and then how compensating perturbations cause continuous decrease in the deviations. That is, the goal towards which he ever strives, is a complete explanation of these complex planetary motions in their totality. Similarly with the geologist. He does not take for his problem only those irregularities of the Earth's crust that are worked by denudation; or only those which igneous

action causes. He does not seek simply to understand
how sedimentary strata were formed; or how faults
were produced; or how moraines originated; or how
the beds of Alpine lakes were scooped out. But taking
into account all agencies co-operating in endless and
ever-varying combinations, he aims to interpret the
entire structure of the Earth's crust. If he studies
separately the actions of rain, rivers, glaciers, icebergs,
tides, waves, volcanoes, earthquakes, etc.; he does so
that he may be better able to comprehend their joint
actions as factors in geological phenomena: the object
of his science being to generalize these phenomena in
all their involved connections, as parts of one whole.
In like manner Biology is the elaboration of a complete theory of Life, in each and all of its involved
manifestations. If different aspects of its phenomena
are investigated apart—if one observer busies himself
in classing organisms, another in dissecting them,
another in ascertaining their chemical compositions,
another in studying functions, another in tracing laws
of modification; they are all, consciously or unconsciously, helping to work out a solution of vital
phenomena in their entirety, both as displayed by
individual organisms and by organisms ·at large.
Thus, in these Concrete Sciences, the object is the
converse of that which the Abstract-Concrete Sciences
propose to themselves. In the one case we have
analytical interpretation; while in the other case we
have *synthetical interpretation.* Instead of synthesis

being used merely to verify analysis; analysis is here used only to aid synthesis. Not to formulate the factors of phenomena is now the object; but to formulate the phenomena resulting from these factors, under the various conditions which the Universe presents.

This third class of Sciences, like the other classes, is divisible into the universal and the non-universal. As there are truths which hold of all phenomena in their elements; so there are truths which hold of all phenomena in their totalities. As force has certain ultimate laws common to its separate modes of manifestation, so in those combinations of its modes which constitute actual phenomena, we find certain ultimate laws that are conformed to in every case. These are the laws of the re-distribution of force. Since we can become conscious of a phenomenon only by some change wrought in us, every phenomenon necessarily implies re-distribution of force—change in the arrangements of matter and motion. Alike in molecular movements and the movements of masses, one great uniformity may be traced. A decreasing quantity of motion, sensible or insensible, always has for its concomitant an increasing aggregation of matter; and, conversely, an increasing quantity of motion, sensible or insensible, has for its concomitant a decreasing aggregation of matter. Give to the molecules of any mass, more of that insensible motion which we call heat, and the parts of the mass become somewhat less closely aggregated. Add a further quantity of insensible motion, and the mass so far disintegrates as to be come liquid. Add still more insensible motion, and the mass disintegrates so completely as to become gas; which occupies a greater space with every extra quantity of insensible motion given to it. On the other hand, every loss of insensible motion by a mass, gaseous, liquid, or solid, is accompanied by a progressing integration of the mass. Similarly with sensible motions, be the bodies moved large or small. Augment the velocities of the planets, and their orbits will enlarge—the Solar System would occupy a wider space. Diminish their velocities, and their orbits will lessen—the Solar System will contract, or become more integrated. And in like manner we see that every sensible motion on the Earth's surface involves a partial disintegration of the moving body from the Earth; while the loss of its motion is accompanied by the body's re-integration with the Earth. In all phenomena we have either an integration of matter and concomitant dissipation of motion; or an absorption of motion and concomitant disintegration of matter. And where, as in living bodies, these processes are going on simultaneously, there is an integration of matter proportioned to the dissipation of motion, and an absorption of motion proportioned to the disintegration of matter. Such, then, are the universal laws of that re-distribution of matter and motion everywhere going on—a re-distribution which results in Evolution so long as the aggregation of matter and dispersion of motion predominate; but which results in Dissolution where there is a predominant aggregation of motion and dispersion of matter. Hence we have a division of Concrete Science which bears towards the other Concrete Sciences, a relation like that which Universal Law of Relation bears to Mathematics, and like that which Universal Mechanics (composition and resolution of forces) bears to Physics. We have a division of Concrete Science which generalizes those concomitants of this re-distribution that hold good among all orders of concrete objects—a division which explains why, along with a predominating integration of matter and dissipation of motion, there goes a change from an indefinite, incoherent homogeneity, to a definite, coherent heterogeneity; and why a reverse re-distribution of matter and motion, is accompanied by a reverse structural change. Passing from this universal Concrete Science, to the non-universal Concrete Sciences; we find that these are primarily divisible into the science which deals with the re-distributions of matter and motion among the masses in space, consequent on their mutual actions as wholes; and the science which deals with the re-distributions of matter and motion consequent on the mutual actions of the molecules in each mass. And of these equally general Sciences, this last is re-divisible into the Science which is limited to the concomitants of re-distribution among the molecules of each mass when regarded as inde-

pendent, and the Science which takes into account the
molecular motion received by radiation from other
masses.   But these sub-divisions, and their sub-sub-
divisions, will be best seen in the annexed Table III.

That these great groups of Sciences and their respective sub-groups, fulfil the definition of a true
classification given at the outset, is, I think, tolerably
manifest.   The subjects of inquiry included in each
primary division, have essential attributes in common
with one another, which they have not in common
with any of the subjects contained in the other primary divisions; and they have, by consequence, a
greater number of common attributes in which they
severally agree with the colligated subjects, and disagree with the subjects otherwise colligated.   Between
Sciences which deal with relations apart from realities,
and Sciences which deal with realities, the distinction is the widest possible; since Being, in some or
all of its attributes, is common to all Sciences of the
second class, and excluded from all Sciences of the first
class.   The distinction between the empty forms of
things and the things themselves, is a distinction
which cannot be exceeded in degree.   And when
we divide the Sciences which treat of realities, into
those which deal with their separate components and
those which deal with their components as united,
we make a profounder distinction than can exist between the Sciences which deal with one or other order

s a predominant integration of Matter and dissipation
n of Matter.

rsc. (*Sidereal Astronomy.*)

n. (*Planetary Astronomy.*)

ind molecules. (*Solar Mineralogy.*)

genesis of radiant forces.*

liquids. (*Solar Meteorology.†*)

rally.

:ion and decomposition of inorganic matters. (*Mineralogy.*)

butions of gases and liquids. (*Meteorology.*)

butions of solids. (*Geology.*)

'phe·
 are { those of structure (*Morphology*) { general. special.

those of function { in their internal relations (*Physiology*) { general. special.

in their exter- nal relations (*Psychology*) { general special { separate. combined. (*Sociology.‡*)

CONCRETE SCIENCE.

{ Universal laws of the continuous re-distribution of Matter and Motion; ·
of Motion, and which results in Dissolution where there is a predominant

Laws of the redistribu-
tions of Matter and Mo-
tion actually going on

{ among the celestial bodies in their
tions to one another as masses: comprel
(ASTRONOMY)

among the molecules
of any celestial
mass; as caused by

{ the actions of these mole
cules on one another
(ASTROGENY)

the actions of these
cules on one another,
with the actions on
of forces radiated by
molecules of other m
(GEOGENY)

TABLE III.

* This must not be supposed to mean chemically-produced forces. The molecular moti
here referred to as dissipated in radiations, is the equivalent of that sensible motion lost dur!
the integration of the mass of molecules, consequent on their mutual gravitation.
+ Embracing the interpretation of such phenomena as the solar spots, the faculæ and t
coronal flames.
‡ Want of space prevents anything beyond the briefest indication of these subdivisions.

h results in Evolution where there is a predominant integration of Matter and dissipation orption of Motion and disintegration of Matter.

:la- { the dynamics of our stellar universe.   (*Sidereal Astronomy.*)
ing { the dynamics of our solar system.   (*Planetary Astronomy.*)

resulting in the formation of compound molecules.   (*Solar Mineralogy.*)

resulting in molecular motions and genesis of radiant forces.*

resulting in movements of gases and liquids.   (*Solar Meteorology.†*)

e- { as exhibited in the planets generally.
·d
m
ıe {
ı :

as exhibited in the Earth {

causing composition and decomposition of inorganic matters.   (*Mineralogy.*)

causing re-distributions of gases and liquids.   (*Meteorology.*)

causing re-distributions of solids.   (*Geology.*)

causing organic phe-nomena; which are (*Biology*) {

those of structure (*Morphology*) { general.
{ special.

those of function {

in their internal relations (*Physiology*) { general.
{ special.

in their exter-nal relations (*Psychology*) { general
{ special { separate.
{ combined.
(*Sociology.‡*)

34

of the components, or than can exist between the Sciences which deal with one or other order of the things composed. The three groups of Sciences may be briefly defined as—laws of the *forms;* laws of the *factors;* laws of the *products.* And when thus defined, it becomes manifest that the groups are so radically unlike in their natures, that there can be no transitions between them; and that any Science belonging to one of the groups must be quite incongruous with the Sciences belonging to either of the other groups, if transferred. How fundamental are the differences between them, will be further seen on considering their functions. The first, or abstract group, is *instrumental* with respect to both the others; and the second, or abstract-concrete group, is *instrumental* with respect to the third or concrete group. An endeavour to invert these functions will at once show how essential is the difference of character. The second and third groups supply subject-matter to the first, and the third supplies subject-matter to the second; but none of the truths which constitute the third group are of any use as solvents of the problems presented by the second group; and none of the truths which the second group formulates can act as solvents of problems contained in the first group. Concerning the sub-divisions of these great groups, little remains to be added. That each of the groups, being co-extensive with all phenomena, contains truths that are universal

and others that are not universal, and that these must
be classed apart, is obvious.   And that the sub-
divisions of the non-universal truths, are to be made in
something like the manner shown in the tables, is
proved by the fact that when the descriptive words
are read from the root to the extremity of any branch,
they form a definition of the Science constituting that
branch.   That the minor divisions might be otherwise arranged, and that better definitions of them
might be given, is highly probable.   They are here
set down merely for the purpose of showing how this
method of classification works out.

I will only further remark, that the relations of the
Sciences as thus represented, are still but imperfectly
represented : their relations cannot be truly shown
on a plane, but only in space of three dimensions.
The three groups cannot rightly be put in linear
order as they have here been.   Since the first stands
related to the third, not only indirectly through the
second, but also directly—it is directly instrumental
with respect to the third, and the third supplies it
directly with subject-matter.   Their relations can
thus only be truly shown by a divergence from a
common root on different sides, in such a way that
each stands in juxta-position to the other two.   And
only by the like mode of arrangement, can the relations
among the sub-divisions of each group be correctly
represented.

IV.

POSTSCRIPT—REPLYING TO CRITICISMS.

POSTSCRIPT, REPLYING TO CRITICISMS.

AMONG objections made to any doctrine, those which come from avowed supporters of an adverse doctrine must be considered, other things equal, as of less weight than those which come from men uncommitted to an adverse doctrine, or but partially committed to it. The element of prepossession, distinctly present in the one case and in the other case mainly or quite absent, is a well-recognized cause of difference in the values of the judgments : supposing the judgments to be otherwise fairly comparable. Hence, when it is needful to bring the replies within a restricted space, a fit course is that of dealing rather with independent criticisms than with criticisms which are really indirect arguments for an opposite view, previously espoused.

For this reason I propose here to confine myself substantially, though not absolutely, to the demurrers entered against the foregoing classification by Prof. Bain, in his recent work on Logic. Before dealing with the more important of these, let me clear the ground by disposing of the less important.

Incidentally, while commenting on the view I take respecting the position of Logic, Prof. Bain points out that this, which is the most abstract of the sciences, owes much to Psychology, which I place among the Concrete Sciences; and he alleges an incongruity between this fact and my statement that the Concrete Sciences are not instrumental in disclosing the truths of the Abstract Sciences. Subsequently he re-raises this apparent anomaly when saying—

> "Nor is it possible to justify the placing of Psychology wholly among Concrete Sciences. It is a highly analytic science, as Mr. Spencer thoroughly knows."

For a full reply, given by implication, I must refer Prof. Bain to § 56 of *The Principles of Psychology*, where I have contended that "while, under its objective aspect, Psychology is to be classed as one of the Concrete Sciences which successively decrease in scope as they increase in speciality; under its subjective aspect, Psychology is a totally unique science, independent of, and antithetically opposed to, all other sciences whatever." A pure idealist will not, I suppose, recognize this distinction; but to every one else it must, I should think, be obvious that the science of subjective existences is the correlative of all the sciences of objective existences; and is as absolutely marked off from them as subject is from object. Objective Psychology, which I class among the Concrete Sciences, is purely synthetic, so long as it is limited, like the other sciences, to objective data; though great aid in the interpretation of these data is derived from the observed correspondence between the phenomena of Objective Psychology as presented in other beings and the phenomena of Subjective Psychology as presented in one's own consciousness. Now it is Subjective Psychology only which is analytic, and which affords aid in the development of Logic. This being explained, the apparent incongruity disappears.

A difficulty raised respecting the manner in which I have expressed the nature of Mathematics, may next be dealt with. Prof. Bain writes:—

> "In the first place, objection may be taken to his language, in discussing the extreme Abstract Sciences, when he speaks of the *empty forms* therein considered. To call Space and Time empty

forms, must mean that they can be thought of without any concrete embodiment whatsoever; that one can think of Time, as a pure abstraction, without having in one's mind any concrete succession. Now, this doctrine is in the last degree questionable."

I quite agree with Prof. Bain that "this doctrine is in the last degree questionable;" but I do not admit that this doctrine is implied by the definition of Abstract Science which I have given. I speak of Space and Time as they are dealt with by mathematicians, and as it is alone possible for pure Mathematics to deal with them. While Mathematics habitually uses in its points, lines, and surfaces, certain existences, it habitually deals with these as representing points, lines, and surfaces that are ideal; and *its conclusions are true only on condition that it does this.* Points having dimensions, lines having breadths, planes having thicknesses, are negatived by its definitions. Using, though it does, material representatives of extension, linear, superficial, or solid, Geometry deliberately ignores their materiality; and attends only to the truths of relation they present. Holding with Prof. Bain, as I do, that our consciousness of Space is disclosed by our experiences of Matter—arguing, as I have done in *The Principles of Psychology*, that it is a consolidated aggregate of all relations of co-existence that have been severally presented by Matter; I nevertheless contend that it is possible to dissociate these relations from Matter to the extent required for formulating them as abstract truths. I contend, too, that this separation is of the kind habitually made in other cases; as, for instance, when the general laws of motion are formulated (as M. Comte's system, among others, formulates them) in such way as to ignore all properties of the bodies dealt with save their powers of taking up, and retaining, and giving out, quantities of motion; though these powers are inconceivable apart from the attribute of extension, which is intentionally disregarded.

Taking other of Prof. Bain's objections, not in the order in which they stand but in the order in which they may be most conveniently dealt with, I quote as follows:—

"The law of the radiation of light (the inverse square of the distance) is said by Mr. Spencer to be Abstract-Concrete, while the disturbing changes in the medium are not to be mentioned except in a Concrete Science of Optics. We need not remark that such a separate handling is unknown to science."

It is perfectly true that "such a separate handling is unknown to science." But, unfortunately for the objection, it is also perfectly true that no such separate handling is proposed by me, or is implied by my classification. How Prof. Bain can have so missed the meaning of the word "concrete," as I have used it, I do not understand. After pointing out that "no one ever drew the line," between the Abstract-Concrete and the Concrete Sciences, "as I have done it," he alleges an anomaly which exists only supposing that I have drawn it where it is ordinarily drawn. He appears inadvertently to have carried with him M. Comte's conception of Optics as a Concrete Science, and, importing it into my classification, debits me with the incongruity. If he will re-read the definition of the Abstract-Concrete Sciences, or study their sub-divisions as shown in Table II., he will, I think, see that the most special laws of the redistribution of light, equally with its most general laws, are included. And if he will pass to the definition and the tabulation of the Concrete Sciences, he will, I think, see no less clearly that Optics cannot be included among them.

Prof. Bain considers that I am not justified in classing Chemistry as an Abstract-Concrete Science, and excluding from it all consideration of the crude forms of the various substances dealt with; and he enforces his dissent by saying that chemists habitually describe the ores and impure mixtures in which the elements, etc., are naturally found. Undoubtedly chemists do this. But do they therefore intend to include an account of the ores of a substance, *as a part of the science* which formulates its molecular constitution and the constitutions of all the definite compounds it enters into? I shall be very much surprised if I find that they do. Chemists habitually prefix to their works a division treating of Molecular Physics; but they do not therefore claim Molecular Physics as a part of Chemistry. If they similarly prefix to the chemistry of each substance an outline of its mineralogy, I do not think they therefore mean to assert that the last belongs to the first. Chemistry proper, embraces nothing beyond an account of the constitutions and modes of action and combining proportions of substances that are taken as absolutely pure; and its truths no more recognize impure substances than the truths of Geometry recognize crooked lines.

Immediately after, in criticizing the fundamental distinction I have made between Chemistry and Biology, as Abstract-Concrete and Concrete respectively, Prof. Bain says:—

"But the objects of Chemistry and the objects of Biology are equally concrete, so far as they go; the simple bodies of chemistry, and their several compounds, are viewed by the Chemist as concrete wholes, and are described by him, not with reference to one factor, but to all their factors."

Issue is here raised in a form convenient for elucidation of the general question. It is true that, *for purposes of identification*, a chemist gives an account of all the sensible characters of a substance. He sets down its crystalline form, its specific gravity, its power of refracting light, its behaviour as magnetic or diamagnetic. But does he thereby include these phenomena as part of the Science of Chemistry? It seems to me that the relation between the weight of any portion of matter and its bulk, which is ascertained on measuring its specific gravity, is a physical and not a chemical fact. I think, too, that the physicist

will claim, as part of his science, all investigations touching
the refraction of light: be the substance producing this
refraction what it may. And the circumstance that the
chemist may test the magnetic or diamagnetic property
of a body, as a means of ascertaining what it is, or as a
means of helping other chemists to determine whether they
have got before them the same body, will neither be held
by the chemist, nor allowed by the physicist, to imply a
transfer of magnetic phenomena from the domain of the
one to that of the other. In brief, though the chemist, in
his account of an element or a compound, may refer to
certain physical traits associated with its molecular constitution and affinities, he does not by so doing change these
into chemical traits. Whatever chemists may put into
their books, Chemistry, considered as a science, includes
only the phenomena of molecular structures and changes—
of compositions and decompositions.* I contend, then,
that Chemistry does *not* give an account of anything
as a concrete whole, in the same way that Biology gives
an account of an organism as a concrete whole. This
will become even more manifest on observing the character
of the biological account. All the attributes of an organism
are comprehended, from the most general to the most special
—from its conspicuous structural traits to its hidden and faint
ones; from its outer actions that thrust themselves on the
attention, to the minutest sub-divisions of its multitudinous

---

* Perhaps some will say that such incidental phenomena as those of the heat
and light evolved during chemical changes, are to be included among chemical
phenomena. I think, however, the physicist will hold that all phenomena of
re-distributed molecular motion, no matter how arising, come within the range
of Physics. But whatever difficulty there may be in drawing the line between
Physics and Chemistry (and, as I have incidentally pointed out in *The Principles
of Psychology*, § 55, the two are closely linked by the phenomena of allotropy
and isomerism), applies equally to the Comtean classification, or to any other.
And I may further point out that no obstacle hence arises to the classification I
am defending. Physics and Chemistry being both grouped by me as Abstract-
Concrete Sciences, no difficulty in satisfactorily dividing them in the least affects
the satisfactoriness of the division of the great group to which they both belong,
from the other two great groups.

internal functions; from its character as a germ, through the many changes of size, form, organization, and habit, it goes through until death; from the physical characters of it as a whole, to the physical characters of its microscopic cells, and vessels, and fibres; from the chemical characters of its substance in general to the chemical characters of each tissue and each secretion—all these, with many others. And not only so, but there is comprehended as the ideal goal of the science, the *consensus* of all these phenomena in their co-existences and successions, as constituting a coherent individualized group definitely combined in space and in time. It is this recognition of *individuality* in its subject-matter, that gives its concreteness to Biology, as to every other Concrete Science. As Astronomy deals with bodies that have their several proper names, or (as with the smaller stars) are registered by their positions, and considers each of them as a distinct individual—as Geology, while dimly perceiving in the Moon and nearest planets other groups of geological phenomena (which it would deal with as independent wholes, did not distance forbid), occupies itself with that individualized group presented by the Earth; so Biology treats either of an individual distinguished from all others, or of parts or products belonging to such an individual, or of structural or functional traits common to many such individuals that have been observed, and supposed to be common to others that are like them in most or all of their attributes. Every biological truth connotes a specifically individualized object, or a number of specifically individualized objects of the same kind, or numbers of different kinds that are severally specific. See, then, the contrast. The truths of the Abstract-Concrete Sciences do not imply specific individuality. Neither Molar Physics, nor Molecular Physics, nor Chemistry, concerns itself with this. The laws of motion are expressed without any reference whatever to the sizes or shapes of the moving masses; which may be taken indifferently to be suns or atoms. The relations between contraction and the escape of molecular motion, and between expansion and the absorption of molecular motion, are expressed in their general forms without reference to the kind of matter; and, if the degree of either that occurs in a particular kind of matter is formulated, no note is taken of the quantity of that matter, much less of its individuality. Similarly with Chemistry. When it inquires into the atomic weight, the molecular structure, the atomicity, the combining proportions, etc., of a substance, it is indifferent whether a grain or a ton be thought of—the conception of amount is absolutely irrelevant. And so with more special attributes. Sulphur, considered chemically, is not sulphur under its crystalline form, or under its allotropic viscid form, or as a liquid, or as a gas; but it is sulphur considered apart from those attributes of quantity, and shape, and state, that give individuality.

Prof. Bain objects to the division I have drawn between the Concrete Science of Astronomy and that Abstract-Concrete Science which deals with the mutually-modified motions of hypothetical masses in space, as "not a little arbitrary." He says:—

"We can suppose a science to confine itself *solely* to the 'factors,' or the separated elements, and never, on any occasion, to combine two into a composite third. This position is intelligible, and possibly defensible. For example, in Astronomy, the Law of Persistence of Motion in a straight line might be discussed in pure ideal separation; and so, the Law of Gravity might be discussed in equally pure separation—both under the Abstract-Concrete department of Mechanics. It might then be reserved to a *concrete* department to unite these in the explanation of a projectile or of a planet. Such, however, is not Mr. Spencer's boundary line. He allows Theoretical Mechanics to make this particular combination, and to arrive at the laws of planetary movement, *in the case of a single planet*. What he does not allow is, to proceed to the case of two planets, mutually disturbing one another, or a planet and a satellite, commonly called the 'problem of the Three Bodies.'"

If I hold what Prof. Bain supposes me to hold, my position would be an absurd one; but he misapprehends me. The misapprehension results in part from his having here, as before, used the word "concrete" with the Comtean meaning, as though it were my meaning; and in part from the inadequacy of my explanation. I did not in the least mean to imply that the Abstract-Concrete Science of Mechanics, when dealing with the motions of bodies in space, is limited to the interpretation of planetary movement such as it would be did only a single planet exist. It never occurred to me that my words (see p. 19) might be so construed. Abstract-Concrete problems admit, in fact, of being complicated indefinitely, without going in the least beyond the definition. I do not draw the line, as Prof. Bain alleges, between the combination of two factors and the combination of three, or between the combination of any number and any greater number. I draw the line between the science which deals with the theory of the factors, taken singly and in combinations of two, three, four, or more, and the science which, *giving to these factors the values derived from observations of actual objects, uses the theory to explain actual phenomena.*

It is true that, in these departments of science, no radical distinction is consistently recognized between theory and the applications of theory. As Prof. Bain says:—

"Newton, in the First Book of the Principia, took up the problem of the Three Bodies, as applied to the Moon, and worked it to exhaustion. So writers on Theoretical Mechanics continue to include the Three Bodies, Precession, and the Tides."

But, supreme though the authority of Newton may be as a mathematician and astronomer, and weighty as are the names of Laplace and Herschel, who in their works have similarly mingled theorems and the explanations yielded by them, it does not seem to me that these facts go for much; unless it can be shown that these writers intended thus to enunciate the views at which they had arrived respecting the classifi-

5

cation of the sciences. Such a union as that presented in their works, adopted merely for the sake of convenience, is, in fact, the indication of incomplete development; and has been paralleled in simpler sciences which have afterwards outgrown it. Two conclusive illustrations are at hand. The name Geometry, utterly inapplicable by its meaning to the science as it now exists, was applicable in that first stage when its few truths were taught in preparation for land-measuring and the setting-out of buildings; but, at a comparatively early date, these comparatively simple truths became separated from their applications, and were embodied by the Greek geometers into systems of theory.* A like purification is now taking place in another division of the science. In the *Géométrie Descriptive* of Monge, theorems were mixed with their applications to projection and plan-drawing. But, since his time, the science and the art have been segregating; and Descriptive Geometry, or, as it may be better termed, the Geometry of Position, is now recognized by mathematicians as a far-reaching system of truths, parts of which are already embodied in books that make no reference to derived methods available by the architect or the engineer. To meet a counter-illustration that will be cited, I may remark that though, in works on Algebra intended for beginners, the theories of quantitative relations, as treated algebraically, are accompanied by groups of problems to be solved, the subject-matters of these problems are not thereby made parts of the Science of Algebra. To say that they are, is to say that Algebra includes the conceptions of distances and relative speeds and times, or of weights and bulks and specific gravities, or of areas ploughed and days and wages; since these, and endless others, may be the terms of

---

* It may be said that the mingling of problems and theorems in Euclid is not quite consistent with this statement; and it is true that we have, in this mingling, a trace of the earlier form of the science. But it is to be remarked that these problems are all purely abstract, and, further, that each of them admits of being expressed as a theorem.

its equations. And just in the same way that these concrete problems, solved by its aid, cannot by any possibility be incorporated with the Abstract Science of Algebra; so I contend that the concrete problems of Astronomy, cannot by any possibility be incorporated with that division of Abstract-Concrete Science which develops the theory of the interactions of free bodies that attract one another.

On this point I find myself at issue, not only with Prof. Bain, but also with Mr. Mill, who contends that :—

"There *is* an abstract science of astronomy, namely, the theory of gravitation, which would equally agree with and explain the facts of a totally different solar system from the one of which our earth forms a part. The actual facts of our own system, the dimensions, distances, velocities, temperatures, physical constitution, etc., of the sun, earth, and planets, are properly the subject of a concrete science, similar to natural history; but the concrete is more inseparably united to the abstract science than in any other case, since the few celestial facts really accessible to us are nearly all required for discovering and proving the law of gravitation as an universal property of bodies, and have therefore an indispensable place in the abstract science as its fundamental data."—*Auguste Comte and Positivism*, p. 43.

In this explanation, Mr. Mill recognizes the fundamental distinction between the Concrete Science of Astronomy, dealing with the bodies actually distributed in space, and a science dealing with hypothetical bodies hypothetically distributed in space. Nevertheless, he regards these sciences as not separable; because the second derives from the first the data whence the law of inter-action is derived. But the truth of this premiss, and the legitimacy of this inference, may alike be questioned. The discovery of the law of inter-action was not due primarily, but only secondarily, to observation of the heavenly bodies. The conception of an inter-acting force that varies inversely as the square of the distance, is an *à priori* conception rationally deducible from mechanical and geometrical considerations. Though unlike in derivation to the many empirical hypotheses of Kepler

respecting planetary orbits and planetary motions, yet it was
like the successful among these in its relation to astronomical
phenomena : it was one of many possible hypotheses, which
admitted of having their consequences worked out and
tested ; and one which, on having its implications compared
with the results of observation, was found to explain them.
In short, the theory of gravitation grew out of experiences
of terrestrial phenomena ; but the verification of it was
reached through experiences of celestial phenomena.   Passing now from premiss to inference, I do not see that, even
were the alleged parentage substantiated, it would necessitate
the supposed inseparability ; any more than the descent of
Geometry from land-measuring necessitates a persistent union
of the two.   In the case of Algebra, as above indicated,
the disclosed laws of quantitative relations hold throughout multitudinous orders of phenomena that are extremely
heterogeneous ; and this makes conspicuous the distinction
between the theory and its applications.   Here the laws of
quantitative relations among masses, distances, velocities, and
momenta, being applied mainly (though not exclusively) to
the concrete cases presented by Astronomy, the distinction
between the theory and its applications is less conspicuous.
But, intrinsically, it is as great in the one case as in the
other.

How great it is, we shall see on taking an analogy.   This
is a living man, of whom we may know little more than that
he is a visible, tangible person ; or of whom we may know
enough to form a voluminous biography.   Again, this book
tells of a fictitious hero, who, like the heroes of old romance,
may be an impersonated virtue or vice, or, like a modern
hero, one of mixed nature, whose various motives and consequent actions are elaborated into a semblance of reality.
But no accuracy and completeness of the picture makes this
fictitious personage an actual personage, or brings him any
nearer to one.   Nor does any meagreness in our knowledge

of a real man reduce him any nearer to the imaginary being of a novel.   To the last, the division between fiction and biography remains an impassable gulf.   So, too, remains the division between the Science dealing with the inter-actions of hypothetical bodies in space, and the Science dealing with the inter-actions of existing bodies in space.   We may elaborate the first to any degree whatever by the introduction of three, four, or any greater number of factors under any number of assumed conditions, until we symbolize a solar system; but to the last an account of our symbolic solar system is as far from an account of the actual solar system as fiction is from biography.

Even more obvious, if it be possible, does the radical character of this distinction become, on observing that from the simplest proposition of General Mechanics we may pass to the most complex proposition of Celestial Mechanics, without a break.   We take a body moving at a uniform velocity, and commence with the proposition that it will continue so to move for ever.   Next, we state the law of its accelerated motion in the same line, when subject to a uniform force. We further complicate the proposition by supposing the force to increase in consequence of approach towards an attracting body; and we may formulate a series of laws of acceleration, resulting from so many assumed laws of increasing attraction (of which the law of gravitation is one). Another factor may now be added by supposing the body to have motion in a direction other than that of the attracting body; and we may determine, according to the ratios of the supposed forces, whether its course will be hyperbolic, parabolic, elliptical, or circular—we may begin with this hypothetical additional force, as infinitesimal, and formulate the varying results as it is little by little increased.   The problem is complicated a degree more by taking into account the effects of a third force, acting in some other direction; and beginning with an infinitesimal amount of this force we may

reach any amount. Similarly, by introducing factor after
factor, each at first insensible in proportion to the rest, we
arrive, through an infinity of gradations, at a combination
of any complexity.

Thus, then, the Science which deals with the inter-action
of hypothetical bodies in space, is *absolutely continuous* with
General Mechanics. We have already seen that it is *absolutely discontinuous* with that account of the heavenly
bodies which has been called Astronomy from the beginning.
When these facts are recognized, it seems to me that there
cannot remain a doubt respecting its true place in a classification of the Sciences.

Passing over minor criticisms, either as met by implication
or as demanding space that cannot be here afforded, let me
say something by way of enforcing the general argument.
I will re-state the case in two ways: the first of them
adapted only to those who accept the general doctrine of
Evolution.

We set out with concentrating nebulous matter. Tracing the re-distributions of this as the rotating contracting
spheroid leaves behind successive annuli, and as these severally breaking up eventually form secondary rotating spheroids,
we come at length to planets in their early stages. Thus
far we consider the phenomena dealt with purely astronomical; and so long as our Earth, regarded as one of
these spheroids, was made up of gaseous and molten
matters only, it presented no definite data for any more
complex Concrete Science. In the lapse of cosmical time
a solid film forms, which, in the course of millions of years,
thickens, and, in the course of further millions of years,
becomes cool enough to permit the precipitation, first of
various other gaseous compounds, and finally of water.
Presently, the varying exposure of different parts of the
spheroid to the Sun's rays, begins to produce appreciable

effects; until at length there have arisen meteorological actions, and consequent geological actions, such as those we now know: determined partly by the Sun's heat, partly by the still-retained internal heat of the Earth, and partly by the action of the Moon on the ocean? How have we reached these geological phenomena? When did the astronomical changes end and the geological begin? It needs but to ask this question to see that there is no real division between the two. Putting pre-conceptions aside, we find nothing more than a group of phenomena continually complicating under the influence of the same original factors; and we see that our conventional division is defensible only on grounds of convenience. Let us advance a stage. As the Earth's surface continues to cool, passing through all degrees of temperature by infinitesimal gradations, the formation of more and more complex inorganic compounds becomes possible; later its surface sinks to that heat at which the less complex compounds of the kinds called organic can exist; and finally the formation of the more complex organic compounds becomes possible. Chemists now show us that these compounds may be built up synthetically in the laboratory—each stage in ascending complexity making possible the next higher stage. Hence it is inferable that, in the myriads of laboratories, endlessly diversified in their materials and conditions, which the Earth's surface furnished during the myriads of years occupied in passing through these stages of temperature, such successive syntheses were effected; and that the highly complex unstable substance out of which all organisms are composed, was eventually formed in microscopic portions: from which, by continuous integrations and differentiations, the evolution of all organisms has proceeded. Where then shall we draw the line between Geology and Biology? The synthesis of this most complex compound, is but a continuation of the syntheses by which all simpler compounds were formed.

The same primary factors have been co-operating with those secondary factors, meteorologic and geologic, previously derived from them. Nowhere do we find a break in the ever-complicating series; for there is a manifest connexion between those movements which various complex compounds undergo during their isomeric transformations, and those changes of form undergone by the protoplasm which we distinguish as living. Strongly contrasted as they eventually become, biological phenomena are at their root inseparable from geological phenomena—inseparable from the aggregate of transformations continually wrought in the matters forming the Earth's surface by the physical forces to which they are exposed. Further stages I need not particularize. The gradual development out of the biological group of phenomena, of the more specialized group we class as psychological, needs no illustration. And when we come to the highest psychological phenomena, it is clear that since aggregations of human beings may be traced upwards from single wandering families to tribes and nations of all sizes and complexities, we pass insensibly from the phenomena of individual human action to those of corporate human action. To resume, then, is it not manifest that in the group of sciences—Astronomy, Geology, Biology, Psychology, Sociology, we have a natural group that admits neither of disruption nor change of order? Here there is both a genetic dependence, and a dependence of interpretations. The phenomena have arisen in this succession in cosmical time; and complete scientific interpretation of each group depends on scientific interpretation of the preceding groups. No other science can be thrust in anywhere without destroying the continuity. To insert Physics between Astronomy and Geology, would be to make a break in the history of a continuous series of changes; and a like break would be produced by inserting Chemistry between Geology and Biology. It is true that Physics and Chemistry are

ELEMENTS OF STATICS AND DYNAMICS. 105

needful as interpreters of these successive assemblages of
facts; but it does not therefore follow that they are themselves to be placed among these assemblages.

Concrete Science, made up of these five concrete sub-
sciences, being thus coherent within itself, and separated
from all other science, there comes the question—Is all other
science similarly coherent within itself? or is it traversed by
some second division that is equally decided? It is thus
traversed. A statical or dynamical theorem, however simple,
has always for its subject-matter something that is conceived.
as extended, and as displaying force or forces—as being a
seat of resistance, or of tension, or of both, and as capable
of possessing more or less of *vis viva*. If we examine the
simplest proposition of Statics, we see that the conception of
Force must be joined with the conception of Space, before
the proposition can be framed in thought; and if we similarly examine the simplest proposition in Dynamics, we see
that Force, Space, and Time, are its essential elements. The
amounts of the terms are indifferent; and, by reduction of
its terms beyond the limits of perception, they are applied to
molecules: Molar Mechanics and Molecular Mechanics are
continuous. From questions concerning the relative motions
of two or more molecules, Molecular Mechanics passes to
changes of aggregation among many molecules, to changes
in the amounts and kinds of the motions possessed by them
as members of an aggregate, and to changes of the motions
transferred through aggregates of them (as those constituting
light). Daily extending its range of interpretations, it is
coming to deal even with the components of each compound
molecule on the same principles. And the unions and dis-
unions of such more or less compound molecules, which
constitute the phenomena of Chemistry, are also being conceived as resultant phenomena of essentially kindred natures
—the affinities of molecules for one another, and their reactions in relation to light, heat, and other modes of force,

being regarded as consequent on the combinations of the
various mechanically-determined motions of their various
components. Without at all out-running, however, this progress in the mechanical interpretation of molecular phenomena, it suffices to point out that the indispensable
elements in any chemical conception are units occupying
places in space, and exerting forces on one another. This,
then, is the common character of all these sciences which
we at present group under the names of Mechanics,
Physics, Chemistry. Leaving undiscussed the question
whether it is possible to conceive of force apart from extended somethings exerting it, we may assert, as beyond
dispute, that if the conception of force be expelled, no
science of Mechanics, Physics, or Chemistry remains. Made
coherent, as these sciences are, by this bond of union, it is
impossible to thrust among them any other science without
breaking their continuity. We cannot place Logic between
Molar Mechanics and Molecular Mechanics. We cannot place
Mathematics between the group of propositions concerning
the behaviour of homogeneous molecules to one another, and
the group of propositions concerning the behaviour of heterogeneous molecules to one another (which we call Chemistry).
Clearly these two sciences lie outside the coherent whole we
have contemplated : separated from it in some radical way.

By what are they radically separated? By the absence of
the conception of force. However true it may be that so
long as Logic and Mathematics have any terms at all, these
must be capable of affecting consciousness, and, by implication, of exerting force ; yet it is the distinctive trait of
these sciences that not only do their propositions make no
reference to such force, but, as far as possible, they deliberately ignore it. Instead of being, as in all the other
sciences, an element that is not only recognized but vital ; in
Mathematics and Logic, force is an element that is not only
not vital, but is studiously not recognized. The terms in

which Logic expresses its propositions, are symbols that do not profess to represent things, properties, or powers, of one kind more than another; and may equally well stand for the attributes belonging to members of some connected series of ideal curves which have never been drawn, as for so many real objects. And the theorems of Geometry, so far from contemplating perceptible lines and surfaces as elements in the truths enunciated, consider these truths as becoming absolute only when such lines and surfaces become ideal— only when the conception of something exercising force is extruded.

Let me now make a second re-statement, not implying acceptance of the doctrine of Evolution, but exhibiting with a clearness almost if not quite as great, these fundamental distinctions.

The concrete sciences, taken together or separately, contemplate as their subject-matters, *aggregates*—either the entire aggregate of sensible existences, or some secondary aggregate separable from this entire aggregate, or some tertiary aggregate separable from this, and so on. Sidereal Astronomy occupies itself with the totality of visible masses distributed through space; which it deals with as made up of identifiable individuals occupying specified places, and severally standing towards one another, towards sub-groups, and towards the entire group, in defined ways. Planetary Astronomy, cutting out of this all-including aggregate that relatively minute part constituting the Solar System, deals with this as a whole—observes, measures, and calculates the sizes, shapes, distances, motions, of its primary, secondary, and tertiary members; and, taking for its larger inquiries the mutual actions of all these members as parts of a co-ordinated assemblage, takes for its smaller inquiries the actions of each member considered as an individual, having a set of intrinsic activities that are modified by a set of extrinsic activities. Restricting itself to one of these aggregates, which admits of close examination, Geology (using this word in its comprehensive meaning) gives an account of terrestrial actions and terrestrial structures, past and present; and, taking for its narrower problems local formations and the agencies to which they are due, takes for its larger problems the serial transformations undergone by the entire Earth. The geologist being occupied with this cosmically small, but otherwise vast, aggregate, the biologist occupies himself with small aggregates formed out of parts of the Earth's superficial substance, and treats each of these as a coordinated whole in its structures and functions; or, when he treats of any particular organ, considers this as a whole made up of parts held in a sub-coordination that refers to the coordination of the entire organism. To the psychologist he leaves those specialized aggregates of functions which adjust the actions of organisms to the complex activities surrounding them: doing this, not simply because they are a stage higher in speciality, but because they are the counterparts of those aggregated states of consciousness dealt with by the science of Subjective Psychology, which stands entirely apart from all other sciences. Finally, the sociologist considers each tribe and nation as an aggregate presenting multitudinous phenomena, simultaneous and successive, that are held together as parts of one combination. Thus, in every case, a concrete science deals with a real aggregate (or a plurality of such aggregates); and it includes as its subject-matter whatever is to be known of this aggregate in respect of its size, shape, motions, density, texture, general arrangement of parts, minute structure, chemical composition, temperature, etc., together with all the multitudinous changes, material and dynamical, gone through by it from the time it begins to exist as an aggregate to the time it ceases to exist as an aggregate.

No abstract-concrete science makes the remotest attempt

to do anything of this sort. Taken together, the abstract-concrete sciences give an account of the various kinds of *properties* which aggregates display; and each abstract-concrete science concerns itself with a certain order of these properties. By this, the properties common to all aggregates are studied and formulated; by that, the properties of aggregates having special forms, special states of aggregation, etc.; and by others, the properties of particular components of aggregates when dissociated from other components. But by all these sciences the aggregate, considered as an individual object, is tacitly ignored; and a property, or a connected set of properties, exclusively occupies attention. It matters not to Mechanics whether the moving mass it considers is a planet or a molecule, a dead stick thrown into the river or the living dog that leaps after it: in any case the curve described by the moving mass conforms to the same laws. Similarly when the physicist takes for his subject the relation between the changing bulk of matter and the changing quantity of molecular motion it contains. Dealing with the subject generally, he leaves out of consideration the kind of matter; and dealing with the subject specially in relation to this or that kind of matter, he ignores the attributes of size and form: save in the still more special cases where the effect on form is considered, and even then size is ignored. So, too, is it with the chemist. A substance he is investigating, never thought of by him as distinguished in extension or amount, is not even required to be perceptible. A portion of carbon on which he is experimenting, may or may not have been visible under its forms of diamond or graphite or charcoal—this is indifferent. He traces it through various disguises and various combinations—now as united with oxygen to form an invisible gas; now as hidden with other elements in such more complex compounds as ether, and sugar, and oil. By sulphuric acid or other agent he precipitates it from these

as a coherent cinder, or as a diffused impalpable powder;
and again, by applying heat, forces it to disclose itself as an
element of animal tissue. Evidently, while thus ascertaining the affinities and atomic equivalence of carbon, the
chemist has nothing to do with any aggregate. He deals
with carbon in the abstract, as something considered apart
from quantity, form, appearance, or temporary state of combination; and conceives it as the possessor of powers or
properties, whence the special phenomena he describes result: the ascertaining of all these powers or properties being
his sole aim.

Finally, the Abstract Sciences ignore alike aggregates and
the powers which aggregates or their components possess;
and occupy themselves with *relations*—either with the relations among aggregates, or among their parts, or the
relations among aggregates and properties, or the relations
among properties, or the relations among relations.   The
same logical formula applies equally well, whether its terms
are men and their deaths, crystals and their planes of cleav-
age, or letters and their sounds.   And how entirely Mathematics concerns itself with relations, we see on remembering
that it has just the same expression for the characters of an
infinitesimal triangle, as for those of the triangle which has
Sirius for its apex and the diameter of the Earth's orbit for
its base.

I cannot see how these definitions of these groups of
sciences can be questioned.   It is undeniable that every
Concrete Science gives an account of an aggregate or of
aggregates, inorganic, organic, or super-organic (a society);
and that, not concerning itself with properties of this or that
order, it concerns itself with the co-ordination of the as-
sembled properties of all orders.   It seems to me no less
certain that an Abstract-Concrete Science gives an account
of some order of properties, general or special; not caring
about the other traits of an aggregate displaying them, and not

recognizing aggregates at all further than is implied by discussion of the particular order of properties. And I think it is equally clear that an Abstract Science, freeing its propositions, so far as the nature of thought permits, from aggregates and properties, occupies itself with the relations of co-existence and sequence, as disentangled from all particular forms of being and action. If then these three groups of sciences are, respectively, accounts of *aggregates*, accounts of *properties*, accounts of *relations*, it is manifest that the divisions between them are not simply perfectly clear, but that the chasms between them are absolute.

Here, perhaps more clearly than before, will be seen the untenability of the classification made by M. Comte. Already (p. 11), after setting forth in a general way these fundamental distinctions, I have pointed out the incongruities that arise when the sciences, conceived as Abstract, Abstract-Concrete, and Concrete, are arranged in the order proposed by him. Such incongruities become still more conspicuous if for these general names of the groups we substitute the definitions given above. The series will then stand thus :—

MATHEMATICS.........An account of *relations*
          (including, under Mechanics, an account of *properties*).
ASTRONOMY ....An account of *aggregates*.
PHYSICS ....An account of *properties*.
CHEMISTRY ....An account of *properties*.
BIOLOGY ....An account of *aggregates*.
SOCIOLOGY ....An account of *aggregates*.

That those who espouse opposite views see clearly the defects in the propositions of their opponents and not those in their own, is a trite remark that holds in philosophical discussions as in all others: the parable of the mote and the beam applies as well to men's appreciations of one another's opinions as to their appreciations of one another's natures. Possibly to my positivist friends I exemplify this truth,—just as they exemplify it to me. Those uncommitted to either view must decide where the mote exists and where the beam. Meanwhile it is clear that one or other of the two views is essentially erroneous; and that no qualifications can bring them into harmony. Either the sciences admit of no such grouping as that which I have described, or they admit of no such serial order as that given by M. Comte.

London,
February, 1871.

V.

REASONS FOR DISSENTING FROM THE PHILOSOPHY OF M. COMTE.

REASONS FOR DISSENTING

FROM THE

PHILOSOPHY OF M. COMTE.

WHILE the preceding pages were passing through the press, there appeared in the *Revue des Deux Mondes* for February 15th, an article on a late work of mine—*First Principles*. To M. Auguste Laugel, the writer of this article, I am much indebted for the careful exposition he has made of some of the leading views set forth in that work; and for the catholic and sympathetic spirit in which he has dealt with them. In one respect, however, M. Laugel conveys to his readers an erroneous impression—an impression doubtless derived from what appears to him adequate evidence, and doubtless expressed in perfect sincerity. M. Laugel describes me as being, in part, a follower of M. Comte. After describing the influence of M. Comte as traceable in the works of some other English writers, naming especially Mr. Mill and Mr. Buckle, he goes on to say that this influence, though not avowed, is easily recognizable in the work he is about to make known; and in several places throughout his review, there are remarks having the same implication. I greatly regret having to take exception to anything said by a critic so candid and so able. But the *Revue des Deux Mondes* circulates widely in England, as well as elsewhere; and finding that there exists in some minds, both here and in America, an impression similar to that entertained by M. Laugel— an impression likely to be confirmed by his statement—it appears to me needful to meet it.

Two causes of quite different kinds, have conspired to diffuse the erroneous belief that M. Comte is an accepted exponent of scientific opinion.   His bitterest foes and his closest friends, have unconsciously joined in propagating it.   On the one hand, M. Comte having designated by the term "Positive Philosophy" all that definitely-established knowledge which men of science have been gradually organizing into a coherent body of doctrine ; and having habitually placed this in opposition to the incoherent body of doctrine defended by theologians ; it has become the habit of the theological party to think of the antagonist scientific party, under the title of "positivists."   And thus, from the habit of calling them "positivists," there has grown up the assumption that they call themselves "positivists," and that they are the disciples of M. Comte.   On the other hand, those who have accepted M. Comte's system, and believe it to be the philosophy of the future, have naturally been prone to see everywhere the signs of its progress ; and wherever they have found opinions in harmony with it, have ascribed these opinions to the influence of its originator.   It is always the tendency of discipleship to magnify the effects of the master's teachings ; and to credit the master with all the doctrines he teaches.   In the minds of his followers, M. Comte's name is associated with scientific thinking, which, in many cases, they first understood from his exposition of it. Influenced as they inevitably are by this association of ideas, they are reminded of M. Comte wherever they meet with thinking which corresponds, in some marked way, to M. Comte's description of scientific thinking ; and hence are apt to imagine him as introducing into other minds, the conceptions which he introduced into their minds.   Such impressions are, however, in most cases quite unwarranted. That M. Comte has given a general exposition of the doctrine and method elaborated by Science, is true.   But it is not true that the holders of this doctrine and followers of this method, are disciples of M. Comte. Neither their modes of inquiry nor their views concerning human knowledge in its nature and limits, are appreciably different from what they were before. If they are "positivists," it is in the sense that all men of science have been more or less consistently "positivists;" and the applicability of M. Comte's title to them, no more makes them his disciples, than does its applicability to men of science who lived and died before M. Comte wrote, make these his disciples. M. Comte himself by no means claims that which some of his adherents are apt, by implication, to claim for him. He says:—"Il y a, sans doute, beaucoup d'analogie entre ma *philosophie positive* et ce que les savans anglais entendent, depuis Newton surtout, par *philosophie naturelle ;*" (see *Avertissement)* and further on he indicates the "grand mouvement imprimé à l'esprit humain, il y a deux siècles, par l'action combinée des préceptes de Bacon, des conceptions de Descartes, et des découvertes de Galilée, comme le moment où l'esprit de la philosophie positive a commencé à se prononcer dans le monde." That is to say, the general mode of thought and way of interpreting phenomena, which M. Comte calls "Positive Philosophy," he recognizes as having been growing for two centuries ; as having reached, when he wrote, a marked development ; and as being the heritage of all men of science.

That which M. Comte proposed to do, was to give scientific thought and method a more definite embodiment and organization ; and to apply it to the interpretation of classes of phenomena not previously dealt with in a scientific manner. The conception was a great one ; and the endeavour to work it out was worthy of sympathy and applause. Some such conception was entertained by Bacon. He, too, aimed at the organization of the sciences ; he, too, held that "Physics is the mother of all the sciences ;" he, too, held that the sciences can be advanced only by combining them,

and saw the nature of the required combination; he, too,
held that moral and civil philosophy could not flourish when
separated from their roots in natural philosophy; and thus
he, too, had some idea of a social science growing out of
physical science. But the state of knowledge in his day prevented any advance beyond the general conception: indeed,
it was marvellous that he should have advanced so far. Instead of a vague, undefined conception, M. Comte has presented the world with a defined and highly-elaborated
conception. In working out this conception he has shown
remarkable breadth of view, great originality, immense fer-
tility of thought, unusual powers of generalization. Considered apart from the question of its truth, his system of
Positive Philosophy is a vast achievement. But after according to M. Comte high admiration for his conception, for
his effort to realize it, and for the faculty he has shown in
the effort to realize it, there remains the inquiry—Has he
succeeded? A thinker who re-organizes the scientific method
and knowledge of his age, and whose re-organization is
accepted by his successors, may rightly be said to have such
successors for his disciples. But successors who accept this
method and knowledge of his age, *minus* his re-organization,
are certainly not his disciples. How then stands the case
with M. Comte? There are some few who receive his
doctrines with but little reservation; and these are his disciples truly so called. There are others who regard with
approval certain of his leading doctrines, but not the rest:
these we may distinguish as partial adherents. There
are others who reject all his distinctive doctrines; and these
must be classed as his antagonists. The members of this
class stand substantially in the same position as they would
have done had he not written. Declining his re-organization of scientific doctrine, they possess this scientific
doctrine in its pre-existing state, as the common heritage
bequeathed by the past to the present; and their adhesion to

this scientific doctrine in no sense implicates them with M. Comte. In this class stand the great body of men of science. And in this case I stand myself.

Coming thus to the personal part of the question, let me first specify those great general principles on which M. Comte is at one with preceding thinkers; and on which I am at one with M. Comte.

All knowledge is from experience, holds M. Comte; and this I also hold—hold it, indeed, in a wider sense than M. Comte: since, not only do I believe that all the ideas acquired by individuals, and consequently all the ideas transmitted by past generations, are thus derived; but I also contend that the very faculties by which they are acquired, are the products of accumulated and organized experiences received by ancestral races of beings (see *Principles of Psychology*). But the doctrine that all knowledge is from experience, is not originated by M. Comte; nor is it claimed by him. He himself says—"Tous les bons esprits répètent, depuis Bacon, qu'il n'y a de connaissances réelle que celles qui reposent sur des faites observés." And the elaboration and definite establishment of this doctrine, has been the special characteristic of the English school of Psychology. Nor am I aware that M. Comte, accepting this doctrine, has done anything to make it more certain, or give it greater definiteness. Indeed it was impossible for him to do so; since he repudiates that part of mental science by which alone this doctrine can be proved.

It is a further belief of M. Comte, that all knowledge is phenomenal or relative; and in this belief I entirely agree. But no one alleges that the relativity of all knowledge was first enunciated by M. Comte. Among others who have more or less consistently held this truth, Sir William Hamilton enumerates, Protagoras, Aristotle, St. Augustin, Boethius, Averroes, Albertus Magnus, Gerson, Leo Hebræus, Melancthon, Scaliger, Francis Piccolomini, Giordano Bruno, Cam-

panella, Bacon, Spinoza, Newton, Kant. And Sir William
Hamilton, in his "Philosophy of the Unconditioned," first
published in 1829, has given a scientific demonstration of this
belief. Receiving it in common with other thinkers, from
preceding thinkers, M. Comte has not, to my knowledge,
advanced this belief. Nor indeed could he advance it, for
the reason already given—he denies the possibility of that
analysis of thought which discloses the relativity of all
cognition.

M. Comte reprobates the interpretation of different classes
of phenomena by assigning metaphysical entities as their
causes; and I coincide in the opinion that the assumption
of such separate entities, though convenient, if not indeed
necessary, for purposes of thought, is, scientifically considered, illegitimate. This opinion is, in fact, a corollary
from the last; and must stand or fall with it. But like the
last it has been held with more or less consistency for generations. M. Comte himself quotes Newton's favorite saying
—"O! Physics, beware of Metaphysics!" Neither to this
doctrine, any more than to the preceding doctrines, has M.
Comte given a firmer basis. He has simply re-asserted it;
and it was out of the question for him to do more. In this
case, as in the others, his denial of subjective psychology
debarred him from proving that these metaphysical entities are
mere symbolic conceptions which do not admit of verification.

Lastly, M. Comte believes in invariable natural laws—
absolute uniformities of relation among phenomena. But
very many before him have believed in them too. Long
familiar even beyond the bounds of the scientific world, the
proposition that there is an unchanging order in things, has,
within the scientific world, held, for generations, the position
of an established postulate: by some men of science recognized only as holding of inorganic phenomena; but recognized by other men of science, as universal. And M. Comte,
accepting this doctrine from the past, has left it substantially

as it was. Though he has asserted new uniformities, I do not think scientific men will admit that he has so *demonstrated* them, as to make the induction more certain; nor has he deductively established the doctrine, by showing that uniformity of relation is a necessary corollary from the persistence of force, as may readily be shown.

These, then, are the pre-established general truths with which M. Comte sets out—truths which cannot be regarded as distinctive of his philosophy. "But why," it will perhaps be asked, "is it needful to point out this; seeing that no instructed reader supposes these truths to be peculiar to M. Comte?" I reply that though no disciple of M. Comte would deliberately claim them for him; and though no theological antagonist at all familiar with science and philophy, supposes M. Comte to be the first propounder of them; yet there is so strong a tendency to associate any doctrines with the name of a conspicuous recent exponent of them, that false impressions are produced, even in spite of better knowledge. Of the need for making this reclamation, definite proof is at hand. In the No. of the *Revue des Deux Mondes* named at the commencement, may be found, on p. 936, the words—" Toute religion, comme toute philosophie, a la prétention de donner une explication de l'univers. La philosophie qui s'appelle *positive* se distingue de toutes les philosophies et de toutes les religions en ce qu'elle a renoncé à cette ambition de l'esprit humain;" and the remainder of the paragraph is devoted to explaining the doctrine of the relativity of knowledge. The next paragraph begins— " Tout imbu de ces idées, que nous exposons sans les discuter pour le moment, M. Spencer divise, etc." Now this is one of those collocations of ideas which tends to create, or to strengthen, the erroneous impression I would dissipate. I do not for a moment suppose that M. Laugel intended to say that these ideas which he describes as ideas of the "Positive Philosophy," are peculiarly the ideas of M. Comte. But

G

little as he probably intended it, his expressions suggest this conception. In the minds of both disciples and antagonists, "the Positive Philosophy" means the philosophy of M. Comte; and to be imbued with the ideas of "the Positive Philosophy" means to be imbued with the ideas of M. Comte —to have received these ideas from M. Comte. After what has been said above, I need scarcely repeat that the conception thus inadvertently suggested, is a wrong one. M. Comte's brief enunciations of these general truths, gave me no clearer apprehensions of them than I had before. Such clarifications of ideas on these ultimate questions, as I can trace to any particular teacher, I owe to Sir William Hamilton.

From the principles which M. Comte held in common with many preceding and contemporary thinkers, let us pass now to the principles that are distinctive of his system. Just as entirely as I agree with M. Comte on those cardinal doctrines which we jointly inherit; so entirely do I disagree with him on those cardinal doctrines which he propounds, and which determine the organization of his philosophy. The best way of showing this will be to compare, side by side, the—

Propositions held by M. Comte.	*Propositions which I hold.*
"... chacune de nos conceptions principales, chaque branche de nos connaissances, passe successivement par trois états théoriques différens: l'état théologique, ou fictif; l'état métaphysique, ou abstrait ; l'état scientifique, ou positif. En d'autres termes, l'esprit humain, par sa nature, emploie successivement dans chacune de ses recherches trois méthodes de philoso-	The progress of our conceptions, and of each branch of knowledge, is from beginning to end intrinsically alike. There are not three methods of philosophizing radically opposed; but one method of philosophizing which remains, in essence, the same. At first, and to the last, the conceived causal agencies of phenomena, have a degree of generality corresponding to the width of the generalizations which experiences have determined; and they change just as gradually as experiences accumulate. The inte-

pher, dont le caractère est essentiellement différent et même radicalement opposé : d'abord la méthode théologique, ensuite la méthode métaphysique, et enfin la méthode positive." p. 3.

gration of causal agencies, originally thought of as multitudinous and local, but finally believed to be one and universal, is a process which involves the passing through all intermediate steps between these extremes; and any appearance of stages can be but superficial. Supposed concrete and individual causal agencies, coalesce in the mind as fast as groups of phenomena are assimilated, or seen to be similarly caused. Along with their coalescence, comes a greater extension of their individualities, and a concomitant loss of distinctness in their individualities. Gradually, by continuance of such coalescences, causal agencies become, in thought, diffused and indefinite. And eventually, without any change in the nature of the process, there is reached the consciousness of a universal causal agency, which cannot be conceived.*

" Le système théologique est parvenu à la plus haute perfection dont il soit susceptible, quand il a substitué l'action providentielle d'un être unique au jeu varié des nombreuses divinités indépendantes qui avaient été imaginées primitivement. De même, le dernier terme du système métaphysique consiste à concevoir, au lieu des différentes entités particulières,

As the progress of thought is one, so is the end one. There are not three possible terminal conceptions ; but only a single terminal conception. When the theological idea of the providential action of one being, is developed to its ultimate form, by the absorption of all independent secondary agencies, it becomes the conception of a being immanent in all phenomena ; and the reduction of it to this state, implies the fading-away, in thought, of all those anthropomorphic attributes by which the aboriginal

* A clear illustration of this process, is furnished by the recent mental integration of Heat, Light, Electricity, etc., as modes of molecular motion. If we go a step back, we see that the modern conception of Electricity, resulted from the integration in consciousness, of the two forms of it evolved in the galvanic battery and in the electric-machine. And going back to a still earlier stage, we see how the conception of statical electricity, arose by the coalescence in thought, of the previously-separate forces manifested in rubbed amber, in rubbed glass, and in lightning. With such illustrations before him, no one can, I think, doubt that the process has been the same from the beginning.

une seule grande entité générale, la *nature*, envisagée comme la source unique de tous les phénomènes. Pareillement, la perfection du système positif, vers laquelle il tend sans cesse, quoiqu'il soit très-probable qu'il ne doive jamais l'atteindre, serait de pouvoir se représenter tous les divers phénomènes observables comme des cas particuliers d'un seul fait général, tel que celui de la gravitation, par exemple." p. 5.

idea was distinguished. The alleged last term of the metaphysical system —the conception of a single great general entity, *nature*, as the source of all phenomena—is a conception identical with the previous one : the consciousness of a single source which, in coming to be regarded as universal, ceases to be regarded as conceivable, differs in nothing but name from the consciousness of one being, manifested in all phenomena. And similarly, that which is described as the ideal state of science—the power to represent all observable phenomena as particular cases of a single general fact, implies the postulating of some ultimate existence of which this single fact is alleged ; and the postulating of this ultimate existence, involves a state of consciousness indistinguishable from the other two.

"...la perfection du système positif, vers laquelle il tend sans cesse, quoiqu'il soit très-probable qu'il ne doive jamais l'atteindre, serait de pouvoir se représenter tous les divers phénomènes observables comme des cas particuliers d'un seul fait général. p. 5 ... ... considérant comme absolument inaccessible, et vide de sens pour nous la recherche de ce qu'on appelle les *causes*, soit premières, soit finales." p. 14.

Though along with the extension of generalizations, and concomitant integration of conceived causal agencies, the conceptions of causal agencies grow more indefinite ; and though as they gradually coalesce into a universal causal agency, they cease to be representable in thought, and are no longer supposed to be comprehensible ; yet the consciousness of *cause* remains as dominant to the last as it was at first ; and can never be got rid of. The consciousness of cause can be abolished only by abolishing consciousness itself.* (*First Principles*, § 26.)

* Possibly it will be said that M. Comte himself admits, that what he calls the perfection of the positive system, will probably never be reached ; and that what he condemns is the inquiry into the *natures* of causes and not the general recognition of cause. To the first of these allegations, I reply that, as I understand M. Comte, the obstacle to the perfect realization of the positive philosophy is the impossibility of carrying generalization so far as to reduce all particular facts to

"Ce n'est pas aux lec- | Ideas do not govern and overthrow

"Ce n'est pas aux lecteurs de cet ouvrage que je croirai jamais devoir prouver que ies idées gouvernent et bouleversent le monde, ou, en d'autres termes, que tout le mécanisme social repose finalement sur des opinions. Ils savent surtout que la grande crise politique et morale des sociétés actuelles tient, en dernière analyse, à l'anarchie intellectuelle." p. 48.*

Ideas do not govern and overthrow the world : the world is governed or overthrown by feelings, to which ideas serve only as guides. The social mechanism does not rest finally upon opinions; but almost wholly upon character. Not intellectual anarchy, but moral antagonism, is the cause of political crises. All social phenomena are produced by the totality of human emotions and beliefs : of which the emotions are mainly pre-determined, while the beliefs are mainly post-determined. Men's desires are chiefly inherited ; but their beliefs are chiefly acquired, and depend on surrounding conditions; and the most important surrounding conditions depend on the social state which the prevalent desires have produced. The social state at any time existing, is the resultant of all the ambitions, self-interests, fears, reverences, indignations, sympathies, etc., of ancestral citizens and existing citizens. The ideas current in this social state, must, on the average, be congruous with the feelings of citizens ; and therefore, on the average, with the social state these feelings have pro-

cases of one general fact—not the impossibility of excluding the consciousness of cause. And to the second allegation I reply, that the essential principle of his philosophy, is an avowed ignoring of cause altogether. For if it is not, *what becomes of his alleged distinction between the perfection of the positive system and the perfection of the metaphysical system ?* And here let me point out that, by affirming exactly the opposite to that which M. Comte thus affirms, I am excluded from the positive school. If his own definition of positivism is to be taken, then, as I hold that what he defines as positivism is an absolute impossibility, it is clear that I cannot be what he calls a positivist.

* A friendly critic alleges that M. Comte is not fairly represented by this quotation, and that he is blamed by his biographer, M. Littré, for his too-great insistance on feeling as a motor of humanity. If in his "Positive Polities," which I presume is here referred to, M. Comte abandons his original position, so much the better. But I am here dealing with what is known as "the Positive Philosophy;" and that the passage above quoted does not misrepresent it, is proved by the fact that this doctrine is re-asserted at the commencement of the Sociology.

duced. Ideas wholly foreign to this social state cannot be evolved, and if introduced from without, cannot get accepted—or, if accepted, die out when the temporary phase of feeling which caused their acceptance, ends. Hence, though advanced ideas when once established, act upon society and aid its further advance; yet the establishment of such ideas depends on the fitness of the society for receiving them. Practically, the popular character and the social state, determine what ideas shall be current; instead of the current ideas determining the social state and the character. The modification of men's moral natures, caused by the continuous discipline of social life, which adapts them more and more to social relations, is therefore the chief proximate cause of social progress. (*Social Statics*, chap. xxx.)

"...je ne dois pas négliger d'indiquer d'avance, comme une propriété essentielle de l'échelle encyclopédique que je vais proposer, sa conformité générale avec l'ensemble de l'histoire scientifique; en ce sens, que, malgré la simultanéité réelle et continue du développement des différentes sciences, celles qui seront classées comme antérieures seront, en effet, plus anciennes et constamment plus avancées que celles présentées comme postérieures." p. 84. . . . .
. . . . "Cet ordre est déterminé par le degré de simplicité, ou, ce qui revient au même, par le degré de généralité des phénomènes." p. 87.

The order in which the generalizations of science are established, is determined by the frequency and impressiveness with which different classes of relations are repeated in conscious experience; and this depends, partly on *the directness with which personal welfare is affected;* partly on *the conspicuousness of one or both the phenomena between which a relation is to be perceived;* partly on *the absolute frequency with which the relations occur;* partly on their *relative frequency of occurrence;* partly on their *degree of simplicity;* and partly on their *degree of abstractness.* (*First Principles*, 1st ed., § 36; appended to this pamphlet.)

"En résultat définitif, la mathématique, l'astronomie, la physique, la chimie, la physiologie, et la physique sociale; telle est la formule enclyopédique qui, parmi le très-grand nombre de classifications que comportent les six sciences fondamentales, est seule logiquement conforme à la hiérarchie naturelle et invariable des phénomènes." p. 115.

The sciences as arranged in this succession specified by M. Comte, *do not* logically conform to the natural and invariable hierarchy of phenomena; and there is no serial order whatever in which they can be placed, which represents either their logical dependence or the dependence of phenomena. (See *Genesis of Science*, and foregoing Essay.)

"On conçoit, en effet, que l'étude rationelle de chaque science fondamentale exigeant la culture préalable de toutes celles qui la précèdent dans notre hiérarchie enclyopédique, n'a pu faire de progrès réels et prendre son véritable caractère, qu' après un grand développement des sciences antérieures relatives à des phénomènes plus généraux, plus abstraits, moins compliqués, et indépendans des autres. C'est donc dans cet ordre que la progression, quoique simultanée, a dû avoir lieu." p. 100.

The historical development of the sciences *has not* taken place in this serial order; nor in any other serial order. There is no "true *filiation* of the sciences." From the beginning, the abstract sciences, the abstract-concrete sciences, and the concrete sciences, have progressed together: the first solving problems which the second and third presented, and growing only by the solution of the problems; and the second similarly growing by joining the first in solving the problems of the third. All along there has been a continuous action and reaction between the three great classes of sciences—an advance from concrete facts to abstract facts, and then an application of such abstract facts to the analysis of new orders of concrete facts. (See *Genesis of Science*.)

Such then are the organizing principles of M. Comte's philosophy. Leaving out of his "*Exposition*" those pre-established general doctrines which are the common property of modern thinkers; these are the general doctrines which remain—these are the doctrines which fundamentally distinguish his system. From every one of them I dissent. . To each proposition I oppose either a widely-different proposition, or a direct negation; and I not only do it now, but have done it from the time when I became acquainted with his writings. This rejection of his cardinal principles should, I think, alone suffice; but there are sundry other views of his, some of them largely characterizing his system, which I equally reject. Let us glance at them.

How organic beings have originated, is an inquiry which M. Comte deprecates as a useless speculation: asserting, as he does, that species are immutable.

This inquiry, I believe, admits of answer, and will be answered. That division of Biology which concerns itself with the origin of species, I hold to be the supreme division, to which all others are subsidiary. For on the verdict of Biology on this matter, must wholly depend our conception of human nature, past, present, and future; our theory of the mind; and our theory of society.

M. Comte contends that of what is commonly known as mental science, all that most important part which consists of the subjective analysis of our ideas, is an impossibility.

I have very emphatically expressed my belief in a subjective science of the mind, by writing a *Principles of Psychology*, one half of which is subjective.

M. Comte's ideal of society is one in which *government* is developed to the greatest extent—in which class-functions are far more under conscious public regulation than now—in which hierarchical organization with unquestioned authority shall guide everything—in which the individual life shall be subordinated in the greatest degree to the social life.

That form of society towards which we are progressing, I hold to be one in which *government* will be reduced to the smallest amount possible, and *freedom* increased to the greatest amount possible — one in which human nature will have become so moulded by social discipline into fitness for the social state, that it will need little external restraint, but will be self-restrained—one in which the citizen will tolerate no interference with his freedom, save that which maintains the equal freedom of others —one in which the spontaneous co-operation which has developed our industrial system, and is now develop·

ing it with increasing rapidity, will produce agencies for the discharge of nearly all social functions, and will leave to the primary govermental agency nothing beyond the function of maintaining those conditions to free action, which make such spontaneous co-operation possible—one in which individual life will thus be pushed to the greatest extent consistent with social life; and in which social life will have no other end than to maintain the completest sphere for individual life

M. Comte, not including in his philosophy the consciousness of a cause manifested to us in all phenomena, and yet holding that there must be a religion, which must have an object, takes for his object —Humanity. "This Collective Life (of Society) is in Comte's system the *Être Suprême;* the only one we can *know,* therefore the only one we can worship."

I conceive, on the other hand, that the object of religious sentiment will ever continue to be, that which it has ever been—the unknown source of things. While the *forms* under which men are conscious of the unknown source of things, may fade away, the *substance* of the consciousness is permanent. Beginning with causal agents conceived as imperfectly known; progressing to causal agents conceived as less known and less knowable; and coming at last to a universal causal agent posited as not to be known at all; the religious sentiment must ever continue to occupy itself with this universal causal agent. Having in the course of evolution, come to have for its object of contemplation, the Infinite Unknowable, the religious sentiment can never again (unless by retrogression) take a Finite Knowable, like Humanity, for its object of contemplation.

Here, then, are sundry other points, all of them important, and the last two supremely important, on which I am diametrically opposed to M. Comte; and did space permit, I could add many others. Radically differing from him as I thus do, in everything distinctive of his philosophy; and having invariably expressed my dissent, publicly and privately, from the time I became acquainted with his writings; it may be imagined that I have been not a little startled to find myself classed as one of the same school. That those who have read *First Principles* only, may have been betrayed into this error in the way above shown, by the ambiguous use of the phrase "Positive Philosophy," I can understand. But that any who are acquainted with my previous writings, should suppose I have any general sympathy with M. Comte, save that implied by preferring proved facts to superstitions, astonishes me.

It is true that, disagreeing with M. Comte, though I do, in all those fundamental views that are peculiar to him, I agree with him in sundry minor views. The doctrine that the education of the individual should accord in mode and arrangement with the education of mankind, considered historically, I have cited from him; and have endeavoured to enforce it. I entirely concur in his opinion that there requires a new order of scientific men, whose function shall be that of co-ordinating the results arrived at by the rest. To him I believe I am indebted for the conception of a social *consensus*; and when the time comes for dealing with this conception, I shall state my indebtedness. And I also adopt his word, Sociology. There are, I believe, in the part of his writings which I have read, various incidental thoughts of great depth and value; and I doubt not that were I to read more of his writings, I should find many others.* It is very probable, too, that I have said (as I am told I have) some things which M. Comte had already said. It would be difficult, I believe, to find any two men who had no opinions in common. And it would be extremely strange if two men,

---

* M. Comte's "Exposition" I read in the original in 1853; and in two or three other places have referred to the original to get his exact words. The Inorganic Physics, and the first chapter of the Biology, I read in Miss Martineau's condensed translation, when it appeared. The rest of M. Comte's views I know only through Mr. Lewes's outline, and through incidental references.

starting from the same general doctrines established by modern science, should traverse some of the same fields of inquiry, without their lines of thought having any points of intersection. But none of these minor agreements can be of much weight in comparison with the fundamental disagreements above specified. Leaving out of view that general community which we both have with the scientific thought of the age, the differences between us are essential, while the correspondences are non-essential. And I venture to think that kinship must be determined by essentials, and not by non-essentials.*

Joined with the ambiguous use of the phrase "Positive Philosophy," which has led to a classing with M. Comte of many men who either ignore or reject his distinctive principles, there has been one special circumstance that has tended to originate and maintain this classing in my own case. The assumption of some relationship between M. Comte and myself, was unavoidably raised by the title of my first book—*Social Statics.* When that book was published, I was unaware that this title had been before used : had I known the fact, I should certainly have adopted an alternative title which I had in view.† If, however, instead of the title,

* In his recent work, *Auguste Comte et la Philosophie Positive*, M. Littré, defending the Comtean classification of the sciences from the criticism I made upon it in the "Genesis of Science," deals with me wholly as an antagonist. The chapter he devotes to his reply, opens by placing me in direct antithesis to the English adherents of Comte, named in the preceding chapter.

† I believed at the time, and have never doubted until now, that the choice of this title was absolutely independent of its previous use by M. Comte. While writing these pages, I have found reason to think the contrary. On referring to *Social Statics*, to see what were my views of social evolution in 1850, when M. Comte was to me but a name, I met with the following sentence :—"Social philosophy may be aptly divided (as political economy has been) into statics and dynamics." (p. 409). This I remembered to be a reference to a division which I had seen in the Political Economy of Mr. Mill. But why had I not mentioned Mr. Mill's name? On referring to the first edition of his work, I found, at the opening of Book iv., this sentence :—"The three preceding parts include as detailed a view as the limits of this treatise permit, of what, by a happy generalization of a mathematical phrase, has been called the Statics of the subject." Here was the solution of the question. The division had not been made by Mr. Mill, but by some writer (on Political Economy I supposed) who was not named by him ; and whom I did not know. It is now manifest, however, that while I supposed I was giving a more extended use to this division, I was but returning to the original use

the work itself be considered, its irrelation to the philosopny of M. Comte, becomes abundantly manifest. There is decisive testimony on this point. In the *North British Review* for August, 1851, a reviewer of *Social Statics* says—

"The title of this work, however, is a complete misnomer. According to all analogy, the phrase "Social Statics" should be used only in some such sense as that in which, as we have already explained, it is used by Comte, namely as designating a branch of inquiry whose end it is to ascertain the laws of social equilibrium or order, as distinct ideally from those of social movement or progress. Of this Mr. Spencer does not seem to have had the slightest notion, but to have chosen the name for his work only as a means of indicating vaguely that it proposed to treat of social concerns in a scientific manner." p. 321.

Respecting M. Comte's application of the words *statics* and *dynamics* to social phenomena, now that I know what it is, I will only say that while I perfectly understand how, by a defensible extension of their mathematical meanings, the one may be used to indicate social *functions in balance*, and the other social *functions out of balance*, I am quite at a loss to understand how the phenomena of *structure* can be included in the one any more than in the other. But the two things which here concern me, are, first, to point out that I had not "the slightest notion" of giving Social Statics the meaning which M. Comte gave it; and, second, to explain the meaning which I did give it. The units of any aggregate of matter, are in equilibrium when they severally act and re-act upon each other on all sides with equal forces. A state of change among them implies that there are forces exercised by some that are not counterbalanced by like forces exercised by others; and a state of rest implies the absence of such uncounterbalanced forces—implies, if the units are homogeneous, equal distances among them— implies a maintenance of their respective spheres of molecular

which Mr. Mill had limited to his special topic. Another thing is, I think, tolerably manifest. As I evidently wished to point out my obligation to some unknown political economist, whose division I thought I was extending, I should have named him had I known who he was. And in that case should not have put this extension of the division as though it were new

motion.   Similarly among the units of a society, the fundamental condition to equilibrium, is, that the restraining forces
which the units exercise on each other, shall be balanced.
If the spheres of action of some units are diminished by
extension of the spheres of action of others, there necessarily
results an unbalanced force which tends to produce political
change in the relations of individuals; and the tendency
to change can cease, only when individuals cease to aggress
on each other's spheres of action—only when there is
maintained that law of equal freedom, which it was the
purpose of *Social Statics* to enforce in all its consequences.
Besides this totally-unlike conception of what constitutes
Social Statics, the work to which I applied that title, is
fundamentally at variance with M. Comte's teachings in
almost everything.   So far from alleging, as M. Comte does,
that society is to be re-organized by philosophy; it alleges
that society is to be re-organized only by the accumulated
effects of habit on character.   Its aim is not the increase
of authoritative control over citizens, but the decrease of it.
A more pronounced individualism, instead of a more pronounced nationalism, is its ideal.   So profoundly is my
political creed at variance with the creed of M. Comte, that,
unless I am misinformed, it has been instanced by a leading
English disciple of M. Comte, as the creed to which he has
the greatest aversion.   One point of coincidence, however,
is recognizable.   The analogy between an individual organism
and a social organism, which was held by Plato and by
Hobbes, is asserted in *Social Statics*, as it is in the *Sociology*
of M. Comte.   Very rightly, M. Comte has made this
analogy the cardinal idea of this division of his philosophy.
In *Social Statics*, the aim of which is essentially ethical,
this analogy is pointed out incidentally, to enforce certain
ethical considerations; and is there obviously suggested
partly by the definition of life which Coleridge derived from
Schelling, and partly by the generalizations of physiologists
there referred to (chap. xxx. §§. 12, 13, 16).   Excepting

this incidental agreement, however, the contents of *Social Statics* are so wholly antagonistic to the philosophy of M. Comte, that, but for the title, the work would never, I think, have raised the remembrance of him—unless, indeed, by the association of opposites.*

And now let me point out that which really *has* exercised a profound influence over my course of thought. The truth which Harvey's embryological inquiries first dimly indicated, which was afterwards more clearly perceived by Wolff, and which was put into a definite shape by Von Baer—the truth that all organic development is a change from a state of homogeneity to a state of heterogeneity—this it is from which very many of the conclusions which I now hold, have indirectly resulted. In *Social Statics*, there is everywhere manifested a dominant belief in the evolution of man and of society. There is also manifested the belief that this evolution is in both cases determined by the incidence of conditions—the actions of circumstances. And there is further, in the sections above referred to, a recognition of the fact that organic and social evolutions, conform to the same law. Falling amid beliefs in evolutions of various orders, everywhere determined by natural causes (beliefs again displayed in the *Theory of Population* and in the *Principles of Psychology*); the formula of Von Baer acted as an organizing principle. The extension of it to other kinds of phenomena than those of individual and social organiza-

---

* Let me add that the conception developed in *Social Statics*, dates back to a series of letters on the "Proper Sphere of Government," published in the *Nonconformist* newspaper, in the latter half of 1842, and republished as a pamphlet in 1843. In these letters will be found, along with many crude ideas, the same belief in the conformity of social phenomena to unvariable laws; the same belief in human progression as determined by such laws; the same belief in the moral modification of men as caused by social discipline; the same belief in the tendency of social arrangements "of themselves to assume a condition of *stable* equilibrium;" the same repudiation of state-control over various departments of social life; the same limitation of state-action to the maintenance of equitable relations among citizens. The writing of *Social Statics* arose from a dissatisfaction with the basis on which the doctrines set forth in those letters were placed: the second half of that work is an elaboration of these doctrines; and the first half a statement of the principles from which they are deducible.

tion, is traceable through successive stages. It may be seen in the last paragraph of an essay on "The Philosophy of Style," published in October, 1852; again in an essay on "Manners and Fashion," published in April, 1854; and then, in a comparatively advanced form, in an essay on "Progess: its Law and Cause," published in April, 1857. Afterwards, there came the recognition of the need for further limitation of this formula; next the inquiry into those general laws of force from which this universal transformation necessarily results; next the deduction of these from the ultimate law of the persistence of force; next the perception that there is everywhere a process of Dissolution complementary to that of Evolution; and, finally, the determination of the conditions (specified in the foregoing essay) under which Evolution and Dissolution respectively occur. The filiation of these results, is, I think, tolerably manifest. The process has been one of continuous development, set up by the addition of Von Baer's law to a number of ideas that were in harmony with it. And I am not conscious of any other influences by which the process has been affected.

It is possible, however, that there may have been influences of which I am not conscious; and my opposition to M. Comte's system may have been one of them. The presentation of antagonistic thoughts, often produces greater definiteness and development of one's own thoughts. It is probable that the doctrines set forth in the essay on "The Genesis of Science," might never have been reached, had not my very decided dissent from M. Comte's conception led me to work them out; and but for this, I might not have arrived at the classification of the sciences exhibited in the foregoing essay. Very possibly there are other cases in which the stimulus of repugnance to M. Comte's views, may have aided in elaborating my own views; though I cannot call to mind any other cases.

Let it by no means be supposed from all I have said, that I do not regard M. Comte's speculations as of great value.

True or untrue, his system as a whole, has doubtless produced important and salutary revolutions of thought in many minds; and will doubtless do so in many more. Doubtless, too, not a few of those who dissent from his general views, have been heathfully stimulated by the consideration of them. The presentation of scientific knowledge and method as a whole, whether rightly or wrongly co-ordinated, cannot have failed greatly to widen the conceptions of most of his readers. And he has done especial service by familiarizing men with the idea of a social science, based on the other sciences. Beyond which benefits resulting from the general character and scope of his philosophy, I believe that there are scattered through his pages, many large ideas that are valuable not only as stimuli, but for their actual truth.

It has been by no means an agreeable task to make these personal explanations; but it has seemed to me a task not to be avoided. Differing so profoundly as I do from M. Comte on all fundamental doctrines, save those which we inherit in common from the past; it has become needful to dissipate the impression that I agree with him—needful to show that a large part of what is currently known as "positive philosophy," is not "positive philosophy" in the sense of being peculiarly M. Comte's philosophy; and to show that beyond that portion of the so-called "positive philosophy" which is not peculiar to him, I dissent from it.

And now at the close, as at the outset, let me express my great regret that these explanations should have been called forth by the statements of a critic who has treated me so liberally. Nothing will, I fear, prevent the foregoing pages from appearing like a very ungracious response to M. Laugel's sympathetically-written review. I can only hope that the gravity of the question at issue, in so far as it concerns myself, may be taken in mitigation, if not as a sufficient apology.

March 12th, 1864.

VI.

OF LAWS IN GENERAL, AND THE ORDER OF THEIR DISCOVERY.

OF LAWS IN GENERAL, AND THE ORDER
OF THEIR DISCOVERY.

[The following chapter was contained in the first edition of First
Principles. *I omitted it from the re-organized second edition, because it did not form an essential part of the new structure. As it is
referred to in the foregoing pages, and as its general argument.is ger-
mane to the contents of those pages, I have thought well to append it
here. Moreover, though I hope eventually to incorporate it in that
division of the* Principles of Sociology *which treats of Intellectual
Progress, yet as it must be long before it can thus re-appear in its
permanent place, and as, should I not get so far in the execution of
my undertaking, it may never thus re-appear at all, it seems proper
to make it more accessible than it is at present. The first and last
sections, which served to link it into the argument of the work to
which it originally belonged, are omitted. The rest has been carefully
revised, and in some parts considerably altered.]*

The recognition of Law being the recognition of uniformity of relations among phenomena, it follows that the
order in which different groups of phenomena are reduced to
law, must depend on the frequency with which the uniform
relations they severally display are distinctly experienced.
At any given stage of progress, those uniformities will be
best known with which men's minds have been oftenest and
most strongly impressed. In proportion partly to the
number of times a relation has been presented to consciousness (not merely to the senses), and in proportion

partly to the vividness with which the terms of the relation
have been cognized, will be the degree in which the constancy of connexion is perceived.

The succession in which relations are generalized being
thus determined, there result certain derivative principles
to which this succession must more immediately and obviously conform.   First is *the directness with which
personal welfare is affected.* While, among surrounding
things, many do not appreciably influence us in any
way, some produce pleasures and some pains, in various
degrees ; and manifestly, those things whose actions on the
organism for good or evil are most decided, will, cœteris
paribus, be those whose laws of action are earliest observed.   Second comes *the conspicuousness of one or both
phenomena between which a relation is to be perceived.* On
every side are phenomena so concealed as to be detected only
by close observation ; others not obtrusive enough to attract
notice ; others which moderately solicit the attention ; others
so imposing or vivid as to force themselves on consciousness ;
and, supposing conditions to be the same, these last will of
course be among the first to have their relations generalized.   In the third place, we have *the absolute frequency
with which the relations occur.* There are coexistences and
sequences of all degrees of commonness, from those which
are ever present to those which are extremely rare ; and
manifestly, the rare coexistences and sequences, as well
as the sequences which are very long in taking place,
will not be reduced to law so soon as those which are
familiar and rapid.   Fourthly has to be added
the relative frequency of occurrence. Many events and appearances are limited to certain times or certain places, or
both ; and, as a relation which does not exist within the
environment of an observer cannot be perceived by him,
however common it may be elsewhere or in another age, we
have to take account of the surrounding physical circum-

stances, as well as of the state of society, of the arts, and of the sciences—all of which affect the frequency with which certain groups of facts are observable. The fifth corollary to be noticed is, that the succession in which different classes of relations are reduced to law, depends in part on their *simplicity*. Phenomena presenting great composition of causes or conditions, have their essential relations so masked, that it requires accumulated experiences to impress upon consciousness the true connexions of antecedents and consequents they involve. Hence, other things equal, the progress of generalization will be from the simple to the complex; and this it is which M. Comte has wrongly asserted to be the sole regulative principle of the progress. Sixth comes *the degree of abstractness*. Concrete relations are the earliest acquisitions. Such analyses of them as separate the essential connexions from their disguising accompaniments, necessarily come later. The analyses of the connexions, always more or less compound, into their elements then becomes possible. And so on continually, until the highest and most abstract truths have been reached.

These, then, are the several derivative principles. The frequency and vividness with which uniform relations are repeated in conscious experience, determining the recognition of their uniformity, and this frequency and vividness depending on the above conditions, it follows that the order in which different classes of facts are generalized, must depend on the extent to which the above conditions are fulfilled in each class. Let us mark how the facts harmonize with this conclusion: taking first a few that elucidate the general truth, and afterwards some that exemplify the special truths which we here see follow from it.

The relations earliest known as uniformities, are those subsisting between the common properties of matter—tangi-

bility, visibility, cohesion, weight, etc. We have no trace of
a time when the resistance offered by an object was regarded
as caused by the will of the object; or when the pressure of
a body on the hand holding it, was ascribed to the agency of a
living being. And accordingly, these are the relations of which
we are oftenest conscious; being objectively frequent, conspicuous, simple, concrete, and of immediate personal concern.

Similarly with the ordinary phenomena of motion. The
fall of a mass on the withdrawal of its support, is a sequence
which directly affects bodily welfare, is conspicuous, simple,
concrete, and very often repeated. Hence it is one of the
uniformities recognized before the dawn of tradition. We
know of no era when movements due to terrestrial gravitation were attributed to volition. Only when the relation
is obscured—only, as in the case of an aërolite, where the
antecedent of the descent is unperceived, do we find the conception of personal agency.   On the other hand, motions of intrinsically the same order as that of a falling stone
—those of the heavenly bodies—long remain ungeneralized;
and until their uniformity is seen, are construed as results of
will. This difference is clearly not dependent on comparative complexity or abstractness; since the motion of a planet
in an ellipse, is as simple and concrete a phenomenon as the
motion of a projected arrow in a parabola. But the antecedents are not conspicuous; the sequences are of long
duration; and they are not often repeated. And that these
are the causes of their slow reduction to law, we see in the
fact that they are severally generalized in the order of their
frequency and conspicuousness—the moon's monthly cycle,
the sun's annual change, the periods of the inferior planets,
the periods of the superior planets.

While astronomical sequences were still ascribed to volition, certain terrestrial sequences of a different kind, but
some of them equally without complication, were interpreted
in like manner. The solidification of water at a low tempe-

rature, is a phenomenon that is simple, concrete, and of
much personal concern. But it is neither so frequent as
those which we see are earliest generalized, nor is the presence of the antecedent so manifest. Though in all but
tropical climates, mid-winter displays the relation between
cold and freezing with tolerable constancy; yet, during the
spring and autumn, the occasional appearance of ice in the
mornings has no very obvious connexion with coldness of
the weather. Sensation being so inaccurate a measure, it is
not possible for the savage to experience the definite relation
between a temperature of 32° and the congealing of water;
and hence the long continued belief in personal agency.
Similarly, but still more clearly, with the winds. The absence of regularity and the inconspicuousness of the antecedents, allowed the mythological explanation to survive for
a great period.

During the era in which the uniformity of many quite
simple inorganic relations was still unrecognized, certain
organic relations, intrinsically very complex and special,
were generalized. The constant coexistence of feathers and
a beak, of four legs with an internal bony framework, are
facts which were, and are, familiar to every savage. Did a
savage find a bird with teeth, or a mammal clothed with
feathers, he would be as much surprised as an instructed
naturalist. Now these uniformities of organic structure thus
early perceived, are of exactly the same kind as those more
numerous ones later established by biology. The constant
coexistence of mammary glands with two occipital condyles
to the skull, of vertebræ with teeth lodged in sockets, of
frontal horns with the habit of rumination, are generalizations as purely empirical as those known to the aboriginal
hunter. The botanist cannot in the least understand the
complex relation between papilionaceous flowers and seeds
borne in flattened pods : he knows these and like connexions
simply in the same way that the barbarian knows the con-

nexions between particular leaves and particular kinds of
wood. But the fact that sundry of the uniform relations
which chiefly make up the organic sciences, were very early
recognized, is due to the high degree of vividness and frequency with which they were presented to consciousness.
Though the connexion between the sounds characteristic of
a bird, and the possession of edible flesh, is extremely involved; yet the two terms of the relation are conspicuous,
often recur in experience, and a knowledge of their connexion has a direct bearing on personal welfare. Meanwhile
innumerable relations of the same order, which are displayed
with even greater frequency by surrounding plants and
animals, remain for thousands of years unrecognised, if they
are unobtrusive or of no apparent moment.

When, passing from this primitive stage to a more advanced stage, we trace the discovery of those less familiar uniformities which mainly constitute what is distinguished as
Science, we find the succession in which knowledge of them
is reached, to be still determined in the same manner. This
will become obvious on contemplating separately the influence of each derivative condition.

How relations that have immediate bearings on the
maintenance of life, are, other things equal, fixed in the
mind before those which have no immediate bearings, the
history of Science abundantly illustrates. The habits of
existing uncivilized races, who fix times by moons and barter
so many of one article for so many of another, show us that
conceptions of equality and number, which are the germs of
mathematical science, were developed under the immediate
pressure of personal wants; and it can scarcely be doubted
that those laws of numerical relations which are embodied in
the rules of arithmetic, were first brought to light through
the practice of mercantile exchange. Similarly with geometry. The derivation of the word shows us that it ori-

ginally included only certain methods of partitioning ground and laying out buildings. The properties of the scales and the lever, involving the first principle in mechanics, were early generalized under the stimulus of commercial and architectural needs. To fix the times of religious festivals and agricultural operations, were the motives which led to the establishment of the simpler astronomic periods. Such small knowledge of chemical relations as was involved in ancient metallurgy, was manifestly obtained in seeking how to improve tools and weapons. In the alchemy of later times, we see how greatly an intense hope of private benefit contributed to the disclosure of a certain class of uniformities. Nor is our own age barren of illustrations. "Here," says Humboldt, when in Guiana, "as in many parts in Europe, the sciences are thought worthy to occupy the mind, only so far as they confer some immediate and practical benefit on society." "How is it possible to believe," said a missionary to him, "that you have left your country to come to be devoured by mosquitoes on this river, and to measure lands that are not your own." · Our coasts furnish like instances. Every sea-side naturalist knows how great is the contempt with which fishermen regard the collection of objects for the microscope or aquarium. Their incredulity as to the possible value of such things is so great, that they can scarcely be induced even by bribes to preserve the refuse of their nets. Nay, we need not go for evidence beyond daily table-talk. The demand for "practical science"—for a knowledge that can be brought to bear on the business of life—joined to the ridicule commonly vented on scientific pursuits having no obvious uses, suffice to show that the order in which laws are discovered greatly depends on the directness with which they affect our welfare.

That, when all other conditions are the same, obtrusive relations will be generalized before unobtrusive ones, is so nearly a truism that examples appear almost superfluous. If

7

it be admitted that by the aboriginal man, as by the child, the co-existent properties of large surrounding objects are noticed before those of minute objects, and that the external relations which bodies present are generalized before their internal relations, it must be admitted that in subsequent stages of progress, the comparative conspicuousness of relations has greatly affected the order in which they were recognized as uniform. Hence it happened that after the establishment of those very manifest sequences constituting a lunation, and those less manifest ones marking a year, and those still less manifest ones marking the planetary periods, astronomy occupied itself with such inconspicuous sequences as those displayed in the repeating cycle of lunar eclipses, and those which suggested the theory of epicycles and eccentrics; while modern astronomy deals with still more inconspicuous sequences, some of which, as the planetary rotations, are nevertheless the simplest which the heavens present. In physics, the early use of canoes implied an empirical knowledge of certain hydrostatic relations that are intrinsically more complex than sundry static relations not empirically known; but these hydrostatic relations were thrust upon observation.   Or, if we compare the solution of the problem of specific gravity by Archimedes with the discovery of atmospheric pressure by Torricelli (the two involving mechanical relations of exactly the same kind), we perceive that the much earlier occurrence of the first than the last was determined, neither by a difference in the irbearings on personal welfare, nor by a difference in the frequency with which illustrations of them came under observation, nor by relative simplicity; but by the greater obtrusiveness of the connexion between antecedent and consequent in the one case than in the other. Among miscellaneous illustrations, it may be pointed out that the connexions between lightning and thunder, and between rain and clouds, were recognized long before others of the same order, simply because they thrust themselves on the attention. Or the long-delayed discovery of the microscopic forms of life, with all the phenomena they present, may be named as very clearly showing how certain groups of relations not ordinarily perceptible, though in other respects like long-familiar relations, have to wait until changed conditions render them perceptible. But, without further details, it needs only to consider the inquiries which now occupy the electrician, the chemist, the physiologist, to see that science has advanced, and is advancing, from the more conspicuous phenomena to the less conspicuous ones.

How the degree of absolute frequency of a relation affects the recognition of its uniformity, we see in contrasting certain biological facts. The connexion between death and bodily injury, constantly displayed not only in men but in all inferior creatures, was known as an instance of natural causation while yet deaths from diseases were thought supernatural. Among diseases themselves, it is observable that unusual ones were regarded as of demoniacal origin during ages when the more frequent were ascribed to ordinary causes: a truth paralleled among our own peasantry, who by the use of charms show a lingering superstition with respect to rare disorders, which they do not show with respect to common ones, such as colds. Passing to physical illustrations, we may note that within the historic period whirlpools were accounted for by the agency of water-spirits; but we do not find that within the same period the disappearance of water on exposure either to the sun or to artificial heat was interpreted in an analogous way: though a more marvellous occurrence, and a much more complex one, its great frequency led to the early recognition of it as a natural uniformity. Rainbows and comets do not differ much in conspicuousness, and a rainbow is intrinsically the more involved phenomenon; but chiefly because of their far greater commonness, rainbows were perceived to have a direct dependence

on sun and rain while yet comets were regarded as signs of
divine wrath.

That races living inland must long have remained ignorant
of the daily and monthly sequences of the tides, and that
tropical races could not early have comprehended the phenomena of northern winters, are extreme illustrations of the
influence which relative frequency has on the recognition of
uniformities.   Animals which, where they are indigenous,
call forth no surprise by their structures or habits, because
these are so familiar, when taken to countries where they
have never been seen, are looked at with an astonishment
approaching to awe—are even thought supernatural: a fact
which will suggest numerous others that show how the localization of phenomena in part controls the order in which they
· are reduced to law. Not only however does their localization
in space affect the progression, but also their localization in
time.   Facts which are rarely if ever manifested in one era,
are rendered very frequent in another, simply through the
changes wrought by civilization.   The lever, of which the
properties are illustrated in the use of sticks and weapons, is
vaguely understood by every savage—on applying it in a
certain way he rightly anticipates certain effects; but the
wheel-and-axle, pulley, and screw, cannot have their powers
either empirically or rationally known till the advance of the
arts has more or less familiarized them.   Through those
various means of exploration which we have inherited and
added to, we have become acquainted with a vast range of
chemical relations that were relatively non-existent to the
primitive man.   To highly-developed industries we owe both
the substances and the appliances that have disclosed to us
countless uniformities which our ancestors had no opportunity of seeing.   These and like instances that will occur
to the reader, show that the accumulated materials, and processes, and products, which characterize the environments of
complex societies, greatly increase the accessibility of various

classes of relations; and by so multiplying the experiences of them, or making them relatively frequent, facilitate their generalization. Moreover, various classes of phenomena presented by society itself, as for instance those which political economy formulates, become relatively frequent, and therefore recognizable, in advanced social states; while in less advanced ones they are either too rarely displayed to have their relations perceived, or, as in the least advanced ones, are not displayed at all.

That, where no other circumstances interfere, the order in which different uniformities are established varies as their complexity, is manifest. The geometry of straight lines was understood before the geometry of curved lines; the properties of the circle before the properties of the ellipse, parabola, and hyperbola; and the equations of curves of single curvature were ascertained before those of curves of double curvature. Plane trigonometry comes in order of time and simplicity before spherical trigonometry; and the mensuration of plane surfaces and solids before the mensuration of curved surfaces and solids. Similarly with mechanics: the laws of simple motion were generalized before those of compound motion; and those of rectilinear motion before those of curvilinear motion. The properties of equal-armed levers or scales, were understood before those of levers with unequal arms; and the law of the inclined plane was formulated earlier than that of the screw, which involves it. In chemistry, the progress has been from the simple inorganic compounds to the more involved or organic compounds. And where, as in the higher sciences, the conditions of the exploration are more complicated, we still may clearly trace relative complexity as determining the order of discovery where other things are equal.

The progression from concrete relations to abstract ones, and from the less abstract to the more abstract, is equally obvious. Numeration, which in its primary form concerned

itself only with groups of actual objects, came earlier than
simple arithmetic ; the rules of which deal with numbers
apart from objects. Arithmetic, limited in its sphere to concrete numerical relations, is alike earlier and less abstract
than Algebra, which deals with the relations of these relations. And in like manner, the Calculus of Operations comes
after Algebra, both in order of evolution and in order of abstractness. In Mechanics, the more concrete relations of
forces exhibited in the lever, inclined plane, etc., were understood before the more abstract relations expressed in the
laws of resolution and composition of forces ; and later than
the three abstract laws of motion as formulated by Newton
came the still more abstract law of inertia. Similarly with
Physics and Chemistry, there has been an advance from
truths entangled in all the specialities of particular facts
and particular classes of facts, to truths disentangled from
the disguising incidents under which they are manifested—
to truths of a higher abstractness.

Brief and rude as is this sketch of a mental development
that has been long and complicated, I venture to think it
shows inductively what was deductively inferred, that the
order in which separate groups of uniformities are recognized, depends not on one circumstance but on several circumstances. The various classes of relations are generalized
in a certain succession, not solely because of one particular
kind of difference in their natures ; but also because they
are variously placed in time and in space, variously open to
observation, and variously related to our own constitutions :
our perception of them being influenced by all these conditions in endless combinations. The comparative degrees
of importance, of obtrusiveness, of absolute frequency, of
relative frequency, of simplicity, of concreteness, are every
one of them factors ; and from their unions in proportions
that are never twice alike, there results a highly complex
process of mental evolution. But while it is thus manifest

that the proximate causes of the succession in which relations
are reduced to law, are numerous and involved; it is also
manifest that there is one ultimate cause to which these
proximate causes are subordinate. As the several circumstances that determine the early or late recognition of uniformities are circumstances that determine the number and
strength of the impressions which these uniformities make
on the mind, it follows that the progression conforms to a
certain fundamental principle of psychology. We see *à
posteriori*, what we concluded *à priori*, that the order in which
relations are generalized, depends on the frequency and
impressiveness with which they are repeated in conscious
experience.

Having roughly analyzed the progress of the past, let
us take advantage of the light thus thrown on the present,
and consider what is implied respecting the future.

Note first that the likelihood of the universality of Law
has been ever growing greater. Out of the countless co-
existences and sequences with which mankind are environed,
they have been continually transferring some from the group
whose order was supposed to be arbitrary, to the group
whose order is known to be uniform. And manifestly, as
fast as the relations that are unreduced to law become
fewer, the probability that among them there are some that
do not conform to law, becomes less. To put the argument
numerically—It is clear that when out of surrounding phenomena a hundred of several kinds have been found to occur
in constant connexions, there arises a slight presumption that
all phenomena occur in constant connexions. When uniformity has been established in a thousand cases, more varied
in their kinds, the presumption gains strength. And when
the known cases of uniformity amount to myriads, including
many of each variety, it becomes an ordinary induction that
uniformity exists everywhere.

152 OF LAWS IN GENERAL.

Silently and insensibly their experiences have been pressing men on towards the conclusion thus drawn. Not out of a conscious regard for these reasons, but from a habit of thought which these reasons formulate and justify, all minds have been advancing towards a belief in the constancy of surrounding coexistences and sequences. Familiarity with concrete uniformities has generated the abstract conception of uniformity—the idea of *Law*; and this idea has been in successive generations slowly gaining fixity and clearness. Especially has it been thus among those whose knowledge of natural phenomena is the most extensive—men of science. The mathematician, the physicist, the astronomer, the chemist, severally acquainted with the vast accumulations of uniformities established by their predecessors, and themselves daily adding new ones as well as verifying the old, acquire a far stronger faith in law than is ordinarily possessed. With them this faith, ceasing to be merely passive, becomes an active stimulus to inquiry. Wherever there exist phenomena of which the dependence is not yet ascertained, these most cultivated intellects, impelled by the conviction that here too there is some invariable connexion, proceed to observe, compare, and experiment; and when they discover the law to which the phenomena conform, as they eventually do, their general belief in the universality of law is further strengthened. So overwhelming is the evidence, and such the effect of this discipline, that to the advanced student of nature, the proposition that there are lawless phenomena has become not only incredible but almost inconceivable.

This habitual recognition of law which already distinguishes modern thought from ancient thought, must spread among men at large. The fulfilment of predictions made possible by every new step, and the further command gained of nature's forces, prove to the uninitiated the validity of scientific generalizations and the doctrine they illustrate. Widening education is daily diffusing among the mass of men that knowledge of these generalizations which has been hitherto confined to the few. And as fast as this diffusion goes on, must the belief of the scientific become the belief of the world at large.

That law is universal, will become an irresistible conclusion when it is perceived that *the progress in the discovery of laws itself conforms to law;* and when this perception makes it clear why certain groups of phenomena have been reduced to law, while other groups are still unreduced. When it is seen that the order in which uniformities are recognized, must depend upon the frequency and vividness with which they are repeated in conscious experience; when it is seen that, as a matter of fact, the most common, important, conspicuous, concrete, and simple, uniformities were the earliest recognized, because they were experienced oftenest and most distinctly; it will by implication be seen that long after the great mass of phenomena have been generalized, there must remain phenomena which, from their rareness, or unobtrusiveness, or seeming unimportance, or complexity, or abstractness, are still ungeneralized. Thus will be furnished a solution to a difficulty sometimes raised. When it is asked why the universality of law is not already fully established, there will be the answer that the directions in which it is not yet established are those in which its establishment must necessarily be latest. That state of things which is inferable beforehand, is just the state which we find to exist. If such coexistences and sequences as those of Biology and Sociology are not yet reduced to law, the presumption is not that they are irreducible to law, but that their laws elude our present means of analysis. Having long ago proved uniformity throughout all the lower classes of relations, and having been step by step proving uniformity throughout classes of relations successively higher and higher, if we have not yet succeeded with the highest classes, it may be fairly concluded that our powers are at fault, rather than that the uniformity does not exist. And unless we make the absurd assumption that the process of generalization, now going on with unexampled rapidity, has reached its limit, and will suddenly cease, we must infer that ul·· timately mankind will discover a constant order of mani· festation even in the most involved and obscure phenomena.

VII.

THE GENESIS OF SCIENCE.

[FROM THE ILLUSTRATIONS OF UNIVERSAL PROGRESS.]

THE GENESIS OF SCIENCE.

THERE has ever prevailed among men a vague notion that scientific knowledge differs in nature from ordinary knowledge. By the Greeks, with whom Mathematics—literally *things learnt*—was alone considered as knowledge proper, the distinction must have been strongly felt; and it has ever since maintained itself in the general mind. Though, considering the contrast between the achievements of science and those of daily unmethodic thinking, it is not surprising that such a distinction has been assumed; yet it needs but to rise a little above the common point of view, to see that no such distinction can really exist: or that at best, it is but a superficial distinction. The same faculties are employed in both cases; and in both cases their mode of operation is fundamentally the same.

If we say that science is organized knowledge, we are met by the truth that all knowledge is organized in a greater or less degree—that the commonest actions of the household and the field presuppose facts colligated, inferences drawn, results expected; and that the general success of these actions proves the data by which they were guided to have been correctly put together. If, again, we say that science is prevision—is a seeing beforehand—is a know-

ing in what times, places, combinations, or sequences, specified phenomena will be found ; we are yet obliged to con
fess that the definition includes much that is utterly foreign
to science in its ordinary acceptation. For example, a child's
knowledge of an apple. This, as far as it goes consists in
previsions. When a child sees a certain form and colours,
it knows that if it puts out its hand it will have certain impressions of resistance, and roundness, and smoothness ;
and if it bites, a certain taste. And manifestly its general
acquaintance with surrounding objects is of like nature—is
made up of facts concerning them, so grouped as that any
part of a group being perceived, the existence of the other
facts included in it is foreseen.

If, once more, we say that science is *exact* prevision, we
still fail to establish the supposed difference. Not only do
we find that much of what we call science is not exact,
and that some of it, as physiology, can never become exact ;
but we find further, that many of the previsions constituting the common stock alike of wise and ignorant, *are* exact. That an unsupported body will fall ; that a lighted candle
will go out when immersed in water ; that ice will melt
when thrown on the fire—these, and many like predictions
relating to the familiar properties of things have as high a
degree of accuracy as predictions are capable of. It is true
that the results predicated are of a very general character ;
but it is none the less true that they are rigorously correct
as far as they go : and this is all that is requisite to fulfil
the definition. There is perfect accordance between the
anticipated phenomena and the actual ones ; and no more
than this can be said of the highest achievements of the
sciences specially characterised as exact.

Seeing thus that the assumed distinction between scientific knowledge and common knowledge is not logically
justifiable ; and yet feeling, as we must, that however impossible it may be to draw a line between them, the two

are not practically identical; there arises the question—What is the relationship that exists between them? A partial answer to this question may be drawn from the illustrations just given. On reconsidering them, it will be observed that those portions of ordinary knowledge which are identical in character with scientific knowledge, comprehend only such combinations of phenomena as are directly cognizable by the senses, and are of simple, invariable nature. That the smoke from a fire which she is lighting will ascend, and that the fire will presently boil water, are provisions which the servant-girl makes equally well with the most learned physicist; they are equally certain, equally exact with his; but they are provisions concerning phenomena in constant and direct relation—phenomena that follow visibly and immediately after their antecedents —phenomena of which the causation is neither remote nor obscure—phenomena which may be predicted by the simplest possible act of reasoning.

If, now, we pass to the previsions constituting what is commonly known as science—that an eclipse of the moon will happen at a specified time; and when a barometer is taken to the top of a mountain of known height, the mercurial column will descend a stated number of inches; that the poles of a galvanic battery immersed in water will give off, the one an inflammable and the other an inflaming gas, in definite ratio—we perceive that the relations involved are not of a kind habitually presented to our senses; that they depend, some of them, upon special combinations of causes; and that in some of them the connection between antecedents and consequents is established only by an elaborate series of inferences. The broad distinction, therefore, between the two orders of knowledge, is not in their nature, but in their remoteness from perception.

If we regard the cases in their most general aspect, we see that the labourer, who, on hearing certain notes in the

adjacent hedge, can describe the particular form and colours of the bird making them ; and the astronomer, who,
having calculated a transit of Venus, can delineate the black
spot entering on the sun's disc, as it will appear through
the telescope, at a specified hour ; do essentially the same
thing. Each knows that on fulfilling the requisite conditions, he shall have a preconceived impression—that after a
definite series of actions will come a group of sensations of
a foreknown kind. The difference, then, is not in the fundamental character of the mental acts ; or in the correctness
of the previsions accomplished by them ; but in the complexity of the processes required to achieve the previsions.
Much of our commonest knowledge is, as far as it goes, rigorously precise. Science does not increase this precision ;
cannot transcend it. What then does it do ? It reduces
other knowledge to the same degree of precision. That
certainty which direct perception gives us respecting coexistences and sequences of the simplest and most accessible kind, science gives us respecting coexistences and sequences, complex in their dependencies or inaccessible to
immediate observation. In brief, regarded from this point
of view, science may be called *an extension of the perceptions by means of reasoning.*

On further considering the matter, however, it will perhaps be felt that this definition does not express the whole
fact—that inseparable as science may be from common
knowledge, and completely as we may fill up the gap between the simplest previsions of the child and the most re-
condite ones of the natural philosopher, by interposing a
series of previsions in which the complexity of reasoning
involved is greater and greater, there is yet a difference
between the two beyond that which is here described. And
this is true. But the difference is still not such as enables
us to draw the assumed line of demarcation. It is a difference not between common knowledge and scientific knowl-

edge; but between the successive phases of science itself, or knowledge itself—whichever we choose to call it. In its earlier phases science attains only to *certainty* of fore-knowledge; in its later phases it further attains to *completeness*. We begin by discovering *a* relation: we end by discovering *the* relation. Our first achievement is to foretell the *kind* of phenomenon which will occur under specific conditions: our last achievement is to foretell not only the kind but the *amount*. Or, to reduce the proposition to its most definite form—undeveloped science is *qualitative* prevision: developed science is *quantitative* prevision.

This will at once be perceived to express the remaining distinction between the lower and the higher stages of positive knowledge. The prediction that a piece of lead will take more force to lift it than a piece of wood of equal size, exhibits certainty, but not completeness, of foresight. The kind of effect in which the one body will exceed the other is foreseen; but not the amount by which it will exceed. There is qualitative prevision only. On the other hand, the prediction that at a stated time two particular planets will be in conjunction; that by means of a lever having arms in a given ratio, a known force will raise just so many pounds; that to decompose a specified quantity of sulphate of iron by carbonate of soda will require so many grains—these predictions exhibit foreknowledge, not only of the nature of the effects to be produced, but of the magnitude, either of the effects themselves, of the agencies producing them, or of the distance in time or space at which they will be produced. There is not only qualitative but quantitative prevision.

And this is the unexpressed difference which leads us to consider certain orders of knowledge as especially scientific when contrasted with knowledge in general. Are the phenomena *measurable?* is the test which we unconsciously

employ. Space is measurable: hence Geometry. Force
and space are measurable: hence Statics. Time, force, and
space are measurable: hence Dynamics. The invention of
the barometer enabled men to extend the principles of mechanics to the atmosphere; and Aerostatics existed. When
a thermometer was devised there arose a science of heat,
which was before impossible. Such of our sensations as we
have not yet found modes of measuring do not originate
sciences. We have no science of smells; nor have we one
of tastes. We have a science of the relations of sounds
differing in pitch, because we have discovered a way to
measure them ; but we have no science of sounds in respect
to their loudness or their *timbre*, because we have got no
measures of loudness and *timbre*.

Obviously it is this reduction of the sensible phenomena
it represents, to relations of magnitude, which gives to any
division of knowledge its especially scientific character.
Originally men's knowledge of weights and forces was in
the same condition as their knowledge of smells and tastes
is now—a knowledge not extending beyond that given by
the unaided sensations ; and it remained so until weighing
instruments and dynamometers were invented. Before
there were hour-glasses and clepsydras, most phenomena
could be estimated as to their durations and intervals, with
no greater precision than degrees of hardness can be estimated by the fingers. Until a thermometric scale was con-
trived, men's judgments respecting relative amounts of
heat stood on the same footing with their present judgments respecting relative amounts of sound. And as in
these initial stages, with no aids to observation, only the
roughest comparisons of cases could be made, and only the
most marked differences perceived ; it is obvious that only
the most simple laws of dependence could be ascertained—
only those laws which being uncomplicated with others,
and not disturbed in their manifestations, required no nice-

ties of observation to disentangle them. Whence it appears not only that in proportion as knowledge becomes quantitative do its previsions become complete as well as certain, but that until its assumption of a quantitative character it is necessarily confined to the most elementary relations.

Moreover it is to be remarked that while, on the one hand, we can discover the laws of the greater proportion of phenomena only by investigating them quantitatively; on the other hand we can extend the range of our quantitative previsions only as fast as we detect the laws of the results we predict. For clearly the ability to specify the magnitude of a result inaccessible to direct measurement, implies knowledge of its mode of dependence on something which can be measured—implies that we know the particular fact dealt with to be an instance of some more general fact. Thus the extent to which our quantitative previsions have been carried in any direction, indicates the depth to which our knowledge reaches in that direction. And here, as another aspect of the same fact, we may further observe that as we pass from qualitative to quantitative prevision, we pass from inductive science to deductive science. Science while purely inductive is purely qualitative: when inaccurately quantitative it usually consists of part induction, part deduction: and it becomes accurately quantitative only when wholly deductive. We do not mean that the deductive and the quantitative are coextensive; for there is manifestly much deduction that is qualitative only. We mean that all quantitative prevision is reached deductively; and that induction can achieve only qualitative prevision.

Still, however, it must not be supposed that these distinctions enable us to separate ordinary knowledge from science; much as they seem to do so. While they show in what consists the broad contrast between the extreme forms of the two, they yet lead us to recognise their essential iden-

tity; and once more prove the difference to be one of degree only. For, on the one hand, the commonest positive
knowledge is to some extent quantitative; seeing that the
amount of the foreseen result is known within certain wide
limits. And, on the other hand, the highest quantitative
prevision does not reach the exact truth, but only a very
near approximation to it. Without clocks the savage
knows that the day is longer in the summer than in the
winter; without scales he knows that stone is heavier than
flesh: that is, he can foresee respecting certain results that
their amounts will exceed these, and be less than those—he
knows *about* what they will be. And, with his most delicate instruments and most elaborate calculations, all that
the man of science can do, is to reduce the difference between the foreseen and the actual results to an unimportant
quantity.

Moreover, it must be borne in mind not only that all
the sciences are qualitative in their first stages,—not only
that some of them, as Chemistry, have but recently reached
the quantitative stage—but that the most advanced sciences
have attained to their present power of determining quantities not present to the senses, or not directly measurable,
by a slow process of improvement extending through thousands of years. So that science and the knowledge of the
uncultured are alike in the nature of their previsions, widely
as they differ in range; they possess a common imperfection, though this is immensely greater in the last than in
the first; and the transition from the one to the other has
been through a series of steps by which the imperfection
has been rendered continually less, and the range continually wider.

These facts, that science and the positive knowledge of
the uncultured cannot be separated in nature, and that the
one is but a perfected and extended form of the other,
must necessarily underlie the whole theory of science, its

progress, and the relations of its parts to each other. There must be serious incompleteness in any history of the sciences, which, leaving out of view the first steps of their genesis, commences with them only when they assume definite forms. There must be grave defects, if not a general untruth, in a philosophy of the sciences considered in their interdependence and development, which neglects the inquiry how they came to be distinct sciences, and how they were severally evolved out of the chaos of primitive ideas.

Not only a direct consideration of the matter, but all analogy, goes to show that in the earlier and simpler stages must be sought the key to all subsequent intricacies. The time was when the anatomy and physiology of the human being were studied by themselves—when the adult man was analyzed and the relations of parts and of functions investigated, without reference either to the relations exhibited in the embryo or to the homologous relations existing in other creatures. Now, however, it has become manifest that no true conceptions, no true generalizations, are possible under such conditions. Anatomists and physiologists now find that the real natures of organs and tissues can be ascertained only by tracing their early evolution ; and that the affinities between existing genera can be satisfactorily made out only by examining the fossil genera to which they are allied. Well, is it not clear that the like must be true concerning all things that undergo development ? Is not science a growth ? Has not science, too, its embryology ? And must not the neglect of its embryology lead to a misunderstanding of the principles of its evolution and of its existing organization ?

There are *à priori* reasons, therefore, for doubting the truth of all philosophies of the sciences which tacitly proceed upon the common notion that scientific knowledge and ordinary knowledge are separate ; instead of commencing, as they should, by affiliating the one upon the

other, and showing how it gradually came to be distinguishable from the other. We may expect to find their
generalizations essentially artificial; and we shall not be
deceived. Some illustrations of this may here be fitly introduced, by way of preliminary to a brief sketch of the
genesis of science from the point of view indicated. And
we cannot more readily find such illustrations than by
glancing at a few of the various *classifications* of the sciences that have from time to time been proposed. To consider all of them would take too much space: we must
content ourselves with some of the latest.

Commencing with those which may be soonest disposed
of, let us notice first the arrangement propounded by Oken
An abstract of it runs thus :—

Part I. MATHESIS.—*Pneumatogeny:* Primary Art, Primary
Consciousness, God, Primary Rest, Time, Polarity, Mo-
tion, Man, Space, Point, Line, Surface, Globe, Rotation.
—*Hylogeny:* Gravity, Matter, Ether, Heavenly Bodies,
Light, Heat, Fire.

(He explains that MATHESIS is the doctrine of the whole;
Pneumatogeny being the doctrine of immaterial totalities, and
Hylogeny that of material totalities.)

Part II. ONTOLOGY.—*Cosmogeny:* Rest, Centre, Motion, Line,
Planets, Form, Planetary System, Comets.—*Stöchiogeny:* Condensation, Simple Matter, Elements, Air,
Water, Earth.—*Stöchiology:* Functions of the Elements,
&c. &c.—*Kingdoms of Nature:* Individuals.

(He says in explanation that "ONTOLOGY teaches us the
phenomena of matter. The first of these are the heavenly
bodies comprehended by *Cosmogeny.* These divide into elements—*Stöchiogeny.* The earth element divides into minerals —*Mineralogy.* These unite into one collective body—
Geogeny. The whole in singulars is the living, or *Organic,*

which again divides into plants and animals. *Biology*, there-fore, divides into *Organogeny, Phytosophy, Zoosophy.*")

FIRST KINGDOM.—MINERALS. *Mineralogy, Geology.*
Part III. BIOLOGY.—*Organosophy, Phytogeny, Phyto-physiology, Phytology, Zoogeny, Physiology, Zoology, Psychology.*'

A glance over this confused scheme shows that it is an attempt to classify knowledge, not after the order in which it has been, or may be, built up in the human conscious-ness; but after an assumed order of creation. It is a pseudo-scientific cosmogony, akin to those which men have enunciated from the earliest times downwards; and only a little more respectable. As such it will not be thought worthy of much consideration by those who, like ourselves, hold that experience is the sole origin of knowledge. Oth-erwise, it might have been needful to dwell on the incon-gruities of the arrangements—to ask how motion can be treated of before space? how there can be rotation with-out matter to rotate? how polarity can be dealt with with-out involving points and lines? But it will serve our pres-ent purpose just to point out a few of the extreme absurdi-ties resulting from the doctrine which Oken seems to hold in common with Hegel, that "to philosophize on Nature is to re-think the great thought of Creation." Here is a sam-ple :—

"Mathematics is the universal science; so also is Phys-io-philosophy, although it is only a part, or rather but a condition of the universe; both are one, or mutually con-gruent.

"Mathematics is, however, a science of mere forms without substance. Physio-philosophy is, therefore, *mathe-matics endowed with substance.*"

From the English point of view it is sufficiently amus-ing to find such a dogma not only gravely stated, but stated as an unquestionable truth. Here we see the experiences of quantitative relations which men have gathered from surrounding bodies and generalized (experiences which had been scarcely at all generalized at the beginning of the historic period)—we find these generalized experiences, these intellectual abstractions, elevated into concrete actualities, projected back into Nature, and considered as the internal frame-work of things—the skeleton by which matter is sustained. But this new form of the old realism, is by no means the most startling of the physio-philosophic principles. We presently read that,

"The highest mathematical idea, or the fundamental principle of all mathematics is the zero = 0." * * *

" Zero is in itself nothing. Mathematics is based upon nothing, and, *consequently*, arises out of nothing.

" Out of nothing, *therefore*, it is possible for something to arise; for mathematics, consisting of propositions, is something, in relation to 0."

By such " consequentlys" and " therefores" it is, that men philosophize when they " re-think the great thought of creation." By dogmas that pretend to be reasons, nothing is made to generate mathematics; and by clothing mathematics with matter, we have the universe! If now we deny, as we *do* deny, that the highest mathematical idea is the zero;—if, on the other hand, we assert, as we *do* assert, that the fundamental idea underlying all mathematics, is that of equality; the whole of Oken's cosmogony disappears. And here, indeed, we may see illustrated, the distinctive peculiarity of the German method of procedure in these matters—the bastard *à priori* method, as it may be termed. The legitimate *à priori* method sets out with propositions of which the negation is inconceivable; the *à priori* method as illegitimately applied, sets out either with propositions of which the negation is *not* inconceivable, or with propositions like Oken's, of which the *affirmation* is inconceivable.

It is needless to proceed further with the analysis; else might we detail the steps by which Oken arrives at the conclusions that "the planets are coagulated colours, for they are coagulated light; that the sphere is the expanded nothing;" that gravity is "a weighty nothing, a heavy essence, striving towards a centre;" that "the earth is the identical, water the indifferent, air the different; or the first the centre, the second the radius, the last the periphery of the general globe or of fire." To comment on them would be nearly as absurd as are the propositions themselves. Let us pass on to another of the German systems of knowledge—that of Hegel.

The simple fact that Hegel puts Jacob Bœhme on a par with Bacon, suffices alone to show that his stand-point is far remote from the one usually regarded as scientific: so far remote, indeed, that it is not easy to find any common basis on which to found a criticism. Those who hold that the mind is moulded into conformity with surrounding things by the agency of surrounding things, are necessarily at a loss how to deal with those, who, like Schelling and Hegel, assert that surrounding things are solidified mind— that Nature is "petrified intelligence." However, let us briefly glance at Hegel's classification. He divides philosophy into three parts:—

1. *Logic*, or the science of the idea in itself, the pure idea.

2. *The Philosophy of Nature*, or the science of the idea considered under its other form—of the idea as Nature.

3. *The Philosophy of the Mind*, or the science of the idea in its return to itself.

Of these, the second is divided into the natural sciences, commonly so called; so that in its more detailed form the series runs thus:—Logic, Mechanics, Physics, Organic Physics, Psychology.

Now, if we believe with Hegel, first, that thought is the

8

true essence of man; second, that thought is the essence of the world; and that, therefore, there is nothing but thought; his classification, beginning with the science of pure thought, may be acceptable. But otherwise, it is an obvious objection to his arrangement, that thought implies things thought of—that there can be no logical forms without the substance of experience—that the science of ideas and the science of things must have a simultaneous origin. Hegel, however, anticipates this objection, and, in his obstinate idealism, replies, that the contrary is true; that all contained in the forms, to become something, requires to be thought: and that logical forms are the foundations of all things.

It is not surprising that, starting from such premises, and reasoning after this fashion, Hegel finds his way to strange conclusions. Out of *space* and *time* he proceeds to build up *motion*, *matter*, *repulsion*, *attraction*, *weight*, and *inertia*. He then goes on to logically evolve the solar system. In doing this he widely diverges from the Newtonian theory; reaches by syllogism the conviction that the planets are the most perfect celestial bodies; and, not being able to bring the stars within his theory, says that they are mere formal existences and not living matter, and that as compared with the solar system they are as little admirable as a cutaneous eruption or a swarm of flies.*

Results so outrageous might be left as self-disproved, were it not that speculators of this class are not alarmed by any amount of incongruity with established beliefs. The only efficient mode of treating systems like this of Hegel, is to show that they are self-destructive—that by their first steps they ignore that authority on which all their subsequent steps depend. If Hegel professes, as he manifestly does, to develop his scheme by reasoning—if he presents

* It is somewhat curious that the author of "The Plurality of Worlds," with quite other aims, should have persuaded himself into similar conclusions.

successive inferences as *necessarily following* from certain premises ; he implies the postulate that a belief which necessarily follows after certain antecedents is a true belief: and, did an opponent reply to one of his inferences, that, though it was impossible to think the opposite, yet the opposite was true, he would consider the reply irrational The procedure, however, which he would thus condemn as destructive of all thinking whatever, is just the procedure exhibited in the enunciation of his own first principles.

Mankind find themselves unable to conceive that there can be thought without things thought of. Hegel, however, asserts that there *can* be thought without things thought of. That ultimate test of a true proposition—the inability of the human mind to conceive the negation of it —which in all other cases he considers valid, he considers invalid where it suits his convenience to do so ; and yet at the same time denies the right of an opponent to follow his example. If it is competent for him to posit dogmas, which are the direct negations of what human consciousness recognises ; then is it also competent for his antagonists to stop him at every step in his argument by saying, that though the particular inference he is drawing seems to his mind, and to all minds, necessarily to follow from the premises, yet it is not true, but the contrary inference is true. Or, to state the dilemma in another form :—If he sets out with inconceivable propositions, then may he with equal propriety make all his succeeding propositions inconceivable ones —may at every step throughout his reasoning draw exactly the opposite conclusion to that which seems involved.

Hegel's mode of procedure being thus essentially suicidal, the Hegelian classification which depends upon it, falls to the ground. Let us consider next that of M. Comte.

As all his readers must admit, M. Comte presents us with a scheme of the sciences which, unlike the foregoing ones, demands respectful consideration. Widely as we differ from him, we cheerfully bear witness to the largeness of his views, the clearness of his reasoning, and the value of his speculations as contributing to intellectual progress. Did we believe a serial arrangement of the sciences to be possible, that of M. Comte would certainly be the one we should adopt. His fundamental propositions are thoroughly intelligible; and if not true, have a great semblance of truth. His successive steps are logically co-ordinated; and he supports his conclusions by a considerable amount of evidence—evidence which, so long as it is not critically examined, or not met by counter evidence, seems to substantiate his positions. But it only needs to assume that antagonistic attitude which *ought* to be assumed towards new doctrines, in the belief that, if true, they will prosper by conquering objectors—it needs but to test his leading doctrines either by other facts than those he cites, or by his own facts differently applied, to at once show that they will not stand. We will proceed thus to deal with the general principle on which he bases his hierarchy of the sciences.

In the second chapter of his *Cours de Philosophie Positive*, M. Comte says:—" Our problem is, then, to find the one *rational* order, amongst a host of possible systems." . . . " This order is determined by the degree of simplicity, or, what comes to the same thing, of generality of their phenomena." And the arrangement he deduces runs thus: *Mathematics, Astronomy, Physics, Chemistry, Physiology, Social Physics.* This he asserts to be " the true *filiation* of the sciences." He asserts further, that the principle of progression from a greater to a less degree of generality, " which gives this order to the whole body of science, arranges the parts of each science." And, finally, he asserts that the gradations thus established *à priori* among the sciences, and the parts of each science, " is

in essential conformity with the order which has sponta
neously taken place among the branches of natural philoso
phy;" or, in other words—corresponds with the order of
historic development.

Let us compare these assertions with the facts. That
there may be perfect fairness, let us make no choice, but
take as the field for our comparison, the succeeding section
treating of the first science—Mathematics; and let us use
none but M. Comte's own facts, and his own admissions.
Confining ourselves to this one science, of course our comparisons must be between its several parts. M. Comte says,
that the parts of each science must be arranged in the
order of their decreasing generality; and that this order
of decreasing generality agrees with the order of historic
development. Our inquiry must be, then, whether the history of mathematics confirms this statement.

Carrying out his principle, M. Comte divides Mathematics into "Abstract Mathematics, or the Calculus (taking the word in its most extended sense) and Concrete
Mathematics, which is composed of General Geometry and
of Rational Mechanics." The subject-matter of the first of
these is *number;* the subject-matter of the second includes
space, time, motion, force. The one possesses the highest
possible degree of generality; for all things whatever
admit of enumeration. The others are less general; seeing that there are endless phenomena that are not cognizable either by general geometry or rational mechanics. In
conformity with the alleged law, therefore, the evolution
of the calculus must throughout have preceded the evolution of the concrete sub-sciences. Now somewhat awk-
wardly for him, the first remark M. Comte makes bearing
upon this point is, that "from an historical point of view,
mathematical analysis *appears to have risen out of* the contemplation of geometrical and mechanical facts." True,
he goes on to say that, "it is not the less independent of

these sciences logically speaking;" for that "analytical
ideas are, above all others, universal, abstract, and simple ·
and geometrical conceptions are necessarily founded on
them."

We will not take advantage of this last passage to
charge M. Comte with teaching, after the fashion of Hegel,
that there can be thought without things thought of. We
are content simply to compare the two assertions, that
analysis arose out of the contemplation of geometrical and
mechanical facts, and that geometrical conceptions are
founded upon analytical ones. Literally interpreted they
exactly cancel each other. Interpreted, however, in a
liberal sense, they imply, what we believe to be de-
monstrable, that the two had *a simultaneous origin.* The
passage is either nonsense, or it is an admission that
abstract and concrete mathematics are coeval. Thus,
at the very first step, the alleged congruity between the
order of generality and the order of evolution, does not
hold good.

But may it not be that though abstract and concrete
mathematics took their rise at the same time, the one
afterwards developed more rapidly than the other; and
has ever since remained in advance of it? No: and again
we call M. Comte himself as witness. Fortunately for his
argument he has said nothing respecting the early stages
of the concrete and abstract divisions after their divergence from a common root; otherwise the advent of
Algebra long after the Greek geometry had reached a high
development, would have been an inconvenient fact for
him to deal with. But passing over this, and limiting
ourselves to his own statements, we find, at the opening of
the next chapter, the admission, that "the historical development of the abstract portion of mathematical science
has, since the time of Descartes, been for the most part
determined by that of the concrete." Further on we read

respecting algebraic functions that "most functions were concrete in their origin—even those which are at present the most purely abstract; and the ancients discovered only through geometrical definitions elementary algebraic properties of functions to which a numerical value was not attached till long afterwards, rendering abstract to us what was concrete to the old geometers." How do these statements tally with his doctrine? Again, having divided the calculus into algebraic and arithmetical, M. Comte admits, as perforce he must, that the algebraic is more general than the arithmetical; yet he will not say that algebra preceded arithmetic in point of time. And again, having divided the calculus of functions into the calculus of direct functions (common algebra) and the calculus of indirect functions (transcendental analysis), he is obliged to speak of this last as possessing a higher generality than the first; yet it is far more modern. Indeed, by implication, M. Comte himself confesses this incongruity; for he says:—"It might seem that the transcendental analysis ought to be studied before the ordinary, as it provides the equations which the other has to resolve; but though the transcendental *is logically independent of the ordinary*, it is best to follow the usual method of study, taking the ordinary first." In all these cases, then, as well as at the close of the section where he predicts that mathematicians will in time "create procedures of *a wider generality*," M. Comte makes admissions that are diametrically opposed to the alleged law.

In the succeeding chapters treating of the concrete department of mathematics, we find similar contradictions. M. Comte himself names the geometry of the ancients *special* geometry, and that of moderns the *general* geometry. He admits that while "the ancients studied geometry with reference to the *bodies* under notice, or specially; the moderns study it with reference to the *phenomena* to be

considered, or generally." He admits that while "the an
cients extracted all they could out of one line or surface
before passing to another," "the moderns, since Descartes,
employ themselves on questions which relate to any figure
whatever." These facts are the reverse of what, according
to his theory, they should be. So, too, in mechanics. Before dividing it into statics and dynamics, M. Comte treats
of the three laws of *motion*, and is obliged to do so; for
statics, the more *general* of the two divisions, though it
does not involve motion, is impossible as a science until the
laws of motion are ascertained. Yet the laws of motion
pertain to dynamics, the more *special* of the divisions.
Further on he points out that after Archimedes, who discovered the law of equilibrium of the lever, statics made
no progress until the establishment of dynamics enabled us
to seek "the conditions of equilibrium through the laws of
the composition of forces." And he adds—"At this day
this is the method universally employed. At the first glance
it does not appear the most rational—dynamics being more
complicated than statics, and precedence being natural to the
simpler. It would, in fact, be more philosophical to refer
dynamics to statics, as has since been done." Sundry discoveries are afterwards detailed, showing how completely
the development of statics has been achieved by considering its problems dynamically; and before the close of the
section M. Comte remarks that "before hydrostatics could
be comprehended under statics, it was necessary that the
abstract theory of equilibrium should be made so general
as to apply directly to fluids as well as solids. This was accomplished when Lagrange supplied, as the basis of the
whole of rational mechanics, the single principle of virtual
velocities." In which statement we have two facts directly
at variance with M. Comte's doctrine;—first, that the simpler science, statics, reached its present development only
by the aid of the principle of virtual velocities, which be-

longs to the more complex science, dynamics ; and that this " single principle " underlying all rational mechanics—this *most general form* which includes alike the relations of statical, hydrostatical, and dynamical forces—was reached so late as the time of Lagrange.

Thus it is *not* true that the historical succession of the divisions of mathematics has corresponded with the order of decreasing generality. It is *not* true that abstract mathematics was evolved antecedently to, and independently of concrete mathematics. It is *not* true that of the subdivisions of abstract mathematics, the more general came before the more special. And it is *not* true that concrete mathematics, in either of its two sections, began with the most abstract and advanced to the less abstract truths.

It may be well to mention, parenthetically, that in defending his alleged law of progression from the general to the special, M. Comte somewhere comments upon the two meanings of the word *general*, and the resulting liability to confusion. Without now discussing whether the asserted distinction can be maintained in other cases, it is manifest that it does not exist here. In sundry of the instances above quoted, the endeavors made by M. Comte himself to disguise, or to explain away, the precedence of the special over the general, clearly indicate that the generality spoken of, is of the kind meant by his formula. And it needs but a brief consideration of the matter to show that, even did he attempt it, he could not distinguish this generality, which, as above proved, frequently comes last, from the generality which he says always comes first. For what is the nature of that mental process by which objects, dimensions, weights, times, and the rest, are found capable of having their relations expressed numerically ? It is the formation of certain abstract conceptions of unity, duality and multiplicity, which are applicable to all things alike. It is the invention of general symbols serving to express the numer ical relations of entities, whatever be their special characters. And what is the nature of the mental process by which numbers are found capable of having their relations expressed algebraically? It is just the same. It is the formation of certain abstract conceptions of numerical func tions which are the same whatever be the magnitudes of the numbers. It is the invention of general symbols serving to express the relations between numbers, as numbers express the relations between things. And transcendental analysis stands to algebra in the same position that algebra stands in to arithmetic.

To briefly illustrate their respective powers;—arithmetic can express in one formula the value of a *particular* tangent to a *particular* curve; algebra can express in one formula the values of *all* tangents to a *particular* curve; transcendental analysis can express in one formula the values of *all* tangents to *all* curves. Just as arithmetic deals with the common properties of lines, areas, bulks, forces, periods; so does algebra deal with the common properties of the numbers which arithmetic presents; so does transcendental analysis deal with the common properties of the equations exhibited by algebra. Thus, the generality of the higher branches of the calculus, when compared with the lower, is the same kind of generality as that of the lower branches when compared with geometry or mechanics. And on examination it will be found that the like relation exists in the various other cases above given.

Having shown that M. Comte's alleged law of progression does not hold among the several parts of the same science, let us see how it agrees with the facts when applied to separate sciences. "Astronomy," says M. Comte, at the opening of Book III., "was a positive science, in its geometrical aspect, from the earliest days of the school of Alexandria; but Physics, which we are now to consider, had no positive character at all till Galileo made his great discov eries on the fall of heavy bodies." On this, our comment is simply that it is a misrepresentation based upon an arbitrary misuse of words—a mere verbal artifice. By choosing to exclude from terrestrial physics those laws of magnitude, motion, and position, which he includes in celestial physics, M. Comte makes it appear that the one owes nothing to the other. Not only is this altogether unwarrantable, but it is radically inconsistent with his own scheme of divisions. At the outset he says—and as the point is important we quote from the original—" Pour la *physique inorganique* nous voyons d'abord, en nous conformant toujours à l'ordre de généralité et de dépendance des phénomènes, qu'elle doit être partagée en deux sections distinctes, suivant qu'elle considère les phénomènes généraux de l'univers, ou, en particulier, ceux que présentent les corps terrestres. D'où la physique céleste, ou l'astronomie, soit géométrique, soit mechanique ; et la physique terrestre."

Here then we have *inorganic physics* clearly divided into *celestial physics* and *terrestrial physics*—the phenomena presented by the universe, and the phenomena presented by earthly bodies. If now celestial bodies and terrestrial bodies exhibit sundry leading phenomena in common, as they do, how can the generalization of these common phenomena be considered as pertaining to the one class rather than to the other ? If inorganic physics includes geometry (which M. Comte has made it do by comprehending *geometrical* astronomy in its sub-section—celestial physics) ; and if its sub-section—terrestrial physics, treats of things having geometrical properties ; how can the laws of geometrical relations be excluded from terrestrial physics ? Clearly if celestial physics includes the geometry of objects in the heavens, terrestrial physics includes the geometry of objects on the earth. And if terrestrial physics includes terrestrial geometry, while celestial physics includes celestial geometry, then the geometrical part of terrestrial physics

precedes the geometrical part of celestial physics; seeing that geometry gained its first ideas from surrounding
objects.   Until men had learnt geometrical relations from
bodies on the earth, it was impossible for them to understand the geometrical relations of bodies in the heavens.

So, too, with celestial mechanics, which had terrestrial
mechanics for its parent.   The very conception of *force*,
which underlies the whole of mechanical astronomy, is bor-
rowed from our earthly experiences; and the leading laws
of mechanical action as exhibited in scales, levers, projec-
tiles, &c., had to be ascertained before the dynamics of the
solar system could be entered upon.   What were the laws
made use of by Newton in working out his grand discovery?
The law of falling bodies disclosed by Galileo; that of the
composition of forces also disclosed by Galileo; and that
of centrifugal force found out by Huyghens—all of them
generalizations of terrestrial physics.   Yet, with facts like
these before him, M. Comte places astronomy before physics in order of evolution!   He does not compare the geometrical parts of the two together, and the mechanical
parts of the two together; for this would by no means
suit his hypothesis.   But he compares the geometrical part
of the one with the mechanical part of the other, and so
gives a semblance of truth to his position.   He is led away
by a verbal delusion.   Had he confined his attention to the
things and disregarded the words, he would have seen that
before mankind scientifically co-ordinated *any one class of
phenomena* displayed in the heavens, they had previously
co-ordinated *a parallel class of phenomena* displayed upon
the surface of the earth.

Were it needful we could fill a score pages with the incongruities of M. Comte's scheme.   But the foregoing samples will suffice.   So far is his law of evolution of the
sciences from being tenable, that, by following his example, and arbitrarily ignoring one class of facts, it would be

possible to present, with great plausibility, just the opposite generalization to that which he enunciates. While he asserts that the rational order of the sciences, like the order of their historic development, "is determined by the degree of simplicity, or, what comes to the same thing, of generality of their phenomena;" it might contrariwise be asserted, that, commencing with the complex and the special, mankind have progressed step by step to a knowledge of greater simplicity and wider generality. So much evidence is there of this as to have drawn from Whewell, in his *History of the Inductive Sciences*, the general remark that "the reader has already seen repeatedly in the course of this history, complex and derivative principles presenting themselves to men's minds before simple and elementary ones."

Even from M. Comte's own work, numerous facts, admissions, and arguments, might be picked out, tending to show this. We have already quoted his words in proof that both abstract and concrete mathematics have progressed towards a higher degree of generality, and that he looks forward to a higher generality still. Just to strengthen this adverse hypothesis, let us take a further instance. From the *particular* case of the scales, the law of equilibrium of which was familiar to the earliest nations known, Archimedes advanced to the more *general* case of the unequal lever with unequal weights; the law of equilibrium of which *includes* that of the scales. By the help of Galileo's discovery concerning the composition of forces, D'Alembert "established, for the first time, the equations of equilibrium of *any* system of forces applied to the different points of a solid body"—equations which include all cases of levers and an infinity of cases besides. Clearly this is progress towards a higher generality—towards a knowledge more independent of special circumstances—towards a study of phenomena "the most disengaged from the incidents of

particular cases;" which is M. Comte's definition of "the
most simple phenomena." Does it not indeed follow from
the familiarly admitted fact, that mental advance is from
the concrete to the abstract, from the particular to the general, that the universal and therefore most simple truths are
the last to be discovered? Is not the government of the
solar system by a force varying inversely as the square of
the distance, a simpler conception than any that preceded
it? Should we ever succeed in reducing all orders of phenomena to some single law—say of atomic action, as M.
Comte suggests—must not that law answer to his test of
being *independent* of all others, and therefore most simple?
And would not such a law generalize the phenomena of
gravity, cohesion, atomic affinity, and electric repulsion, just
as the laws of number generalize the quantitative phenomena of space, time and force?

The possibility of saying so much in support of an hypothesis the very reverse of M. Comte's, at once proves that
his generalization is only a half-truth. The fact is, that
neither proposition is correct by itself; and the actuality is
expressed only by putting the two together. The progress
of science is duplex: it is at once from the special to the
general, and from the general to the special: it is analytical
and synthetical at the same time.

M. Comte himself observes that the evolution of science
has been accomplished by the division of labour; but he
quite misstates the mode in which this division of labour
has operated. As he describes it, it has simply been an arrangement of phenomena into classes, and the study of each
class by itself. He does not recognise the constant effect
of progress in each class upon *all* other classes; but only on
the class succeeding it in his hierarchical scale. Or if he
occasionally admits collateral influences and intercommuni-
cations, he does it so grudgingly, and so quickly puts the
admissions out of sight and forgets them, as to leave the

impression that, with but trifling exceptions, the sciences aid each other only in the order of their alleged succession. The fact is, however, that the division of labour in science, like the division of labour in society, and like the "physiological division of labour" in individual organisms, has been not only a specialization of functions, but a continuous helping of each division by all the others, and of all by each. Every particular class of inquirers has, as it were, secreted its own particular order of truths from the general mass of material which observation accumulates; and all other classes of inquirers have made use of these truths as fast as they were elaborated, with the effect of enabling them the better to elaborate each its own order of truths.

It was thus in sundry of the cases we have quoted as at variance with M. Comte's doctrine. It was thus with the application of Huyghens's optical discovery to astronomical observation by Galileo. It was thus with the application of the isochronism of the pendulum to the making of instruments for measuring intervals, astronomical and other. It was thus when the discovery that the refraction and dispersion of light did not follow the same law of variation, affected both astronomy and physiology by giving us achromatic telescopes and microscopes. It was thus when Bradley's discovery of the aberration of light enabled him to make the first step towards ascertaining the motions of the stars. It was thus when Cavendish's torsion-balance experiment determined the specific gravity of the earth, and so gave a datum for calculating the specific gravities of the sun and planets. It was thus when tables of atmospheric refraction enabled observers to write down the real places of the heavenly bodies instead of their apparent places. It was thus when the discovery of the different expansibilities of metals by heat, gave us the means of correcting our chronometrical measurements of astronomical periods. It was thus when the lines of the prismatic spectrum were

used to distinguish the heavenly bodies that are of like nature with the sun from those which are not.   It was thus
when, as recently, an electro-telegraphic instrument was invented for the more accurate registration of meridional
transits.   It was thus when the difference in the rates of a
clock at the equator, and nearer the poles, gave data for
calculating the oblateness of the earth, and accounting for
the precession of the equinoxes.   It was thus—but it is
needless to continue.

   Here, within our own limited knowledge of its history, we
have named ten additional cases in which the single science
of astronomy has owed its advance to sciences coming *after*
it in M. Comte's series.   Not only its secondary steps, but
its greatest revolutions have been thus determined.   Kepler could not have discovered his celebrated laws had it not
been for Tycho Brahe's accurate observations ; and it was
only after some progress in physical and chemical science
that the improved instruments with which those observations were made, became possible.   The heliocentric theory
of the solar system had to wait until the invention of the
telescope before it could be finally established.   Nay, even
the grand discovery of all—the law of gravitation—depended for its proof upon an operation of physical science, the
measurement of a degree on the Earth's surface.   So completely indeed did it thus depend, that Newton *had actually
abandoned his hypothesis* because the length of a degree,
as then stated, brought out wrong results ; and it was only
after Picart's more exact measurement was published, that
he returned to his calculations and proved his great generalization.   Now this constant intercommunion, which, for
brevity's sake, we have illustrated in the case of one science
only, has been taking place with all the sciences.   Throughout the whole course of their evolution there has been a
continuous *consensus* of the sciences—a *consensus* exhibiting a general correspondence with the *consensus* of facul

ties in each phase of mental development; the one being an objective registry of the subjective state of the other.

From our present point of view, then, it becomes obvious that the conception of a *serial* arrangement of the sciences is a vicious one. It is not simply that the schemes we have examined are untenable; but it is that the sciences cannot be rightly placed in any linear order whatever. It is not simply that, as M. Comte admits, a classification "will always involve something, if not arbitrary, at least artificial;" it is not, as he would have us believe, that, neglecting minor imperfections a classification may be substantially true; but it is that any grouping of the sciences in a succession gives a radically erroneous idea of their genesis and their dependencies. There is no "one *rational* order among a host of possible systems." There is no "true *filiation* of the sciences." The whole hypothesis is fundamentally false. Indeed, it needs but a glance at its origin to see at once how baseless it is. Why a *series?* What reason have we to suppose that the sciences admit of a *linear* arrangement? Where is our warrant for assuming that there is some *succession* in which they can be placed? There is no reason; no warrant. Whence then has arisen the supposition? To use M. Comte's own phraseology, we should say, it is a metaphysical conception. It adds another to the cases constantly occurring, of the human mind being made the measure of Nature. We are obliged to think in sequence; it is the law of our minds that we must consider subjects separately, one after another: *therefore* Nature must be serial—*therefore* the sciences must be classifiable in a succession. See here the birth of the notion, and the sole evidence of its truth. Men have been obliged when arranging in books their schemes of education and systems of knowledge, to choose *some* order or other. And from inquiring what is the best order, have naturally fallen into the belief that there is an order which truly represents the facts—have persevered in seeking such an order; quite overlooking the previous question whether it is likely that Nature has consulted the convenience of book-making.

For German philosophers, who hold that Nature is "petrified intelligence," and that logical forms are the foundations of all things, it is a consistent hypothesis that as thought is serial, Nature is serial; but that M. Comte, who is so bitter an opponent of all anthropomorphism, even in its most evanescent shapes, should have committed the mistake of imposing upon the external world an arrangement which so obviously springs from a limitation of the human consciousness, is somewhat strange. And it is the more strange when we call to mind how, at the outset, M. Comte remarks that in the beginning "*toutes les sciences sont cultivées simultanément par les mêmes esprits ;*" that this is "*inevitable et même indispensable ;*" and how he further remarks that the different sciences are "*comme les diverses branches d'un tronc unique.*" Were it not accounted for by the distorting influence of a cherished hypothesis, it would be scarcely possible to understand how, after recognising truths like these, M. Comte should have persisted in attempting to construct "*une échelle encyclopédique.*"

The metaphor which M. Comte has here so inconsistently used to express the relations of the sciences—branches of one trunk—is an approximation to the truth, though not the truth itself. It suggests the facts that the sciences had a common origin; that they have been developing simultaneously; and that they have been from time to time dividing and sub-dividing. But it does not suggest the yet more important fact, that the divisions and sub-divisions thus arising do not remain separate, but now and again re-unite in direct and indirect ways. They

inosculate ; they severally send off and receive connecting
growths ; and the intercommunion has been ever becom·
ing more frequent, more intricate, more widely ramified.
There has all along been higher specialization, that there
might be a larger generalization ; and a deeper analysis,
that there might be a better synthesis. Each larger generalization has lifted sundry specializations still higher ; and
each better synthesis has prepared the way for still deeper
analysis.

And here we may fitly enter upon the task awhile since
indicated—a sketch of the Genesis of Science, regarded as
a gradual outgrowth from common knowledge—an extension of the perceptions by the aid of the reason. We propose to treat it as a psychological process historically displayed ; tracing at the same time the advance from qualitative to quantitative prevision ; the progress from concrete
facts to abstract facts, and the application of such abstract
facts to the analysis of new orders of concrete facts ; the
simultaneous advance in gereralization and specialization ;
the continually increasing subdivision and reunion of the
sciences ; and their constantly improving *consensus*.

To trace out scientific evolution from its deepest roots
would, of course, involve a complete analysis of the mind.
For as science is a development of that common knowledge
acquired by the unaided senses and uncultured reason, so
is that common knowledge itself gradually built up out of
the simplest perceptions. We must, therefore, begin
somewhere abruptly ; and the most appropriate stage
to take for our point of departure will be the adult mind
of the savage.

Commencing thus, without a proper preliminary analysis, we are naturally somewhat at a loss how to present, in
a satisfactory manner, those fundamental processes of
thought out of which science ultimately originates. Per·

haps our argument may be best initiated by the proposi
tion, that all intelligent action whatever depends upon the
discerning of distinctions among surrounding things.   The
condition under which only it is possible for any creature
to obtain food and avoid danger is, that it shall be differently affected by different objects—that it shall be led to
act in one way by one object, and in another way by
another.   In the lower orders of creatures this condition is
fulfilled by means of an apparatus which acts automatically.
In the higher orders the actions are partly automatic,
partly conscious.   And in man they are almost wholly
conscious.

Throughout, however, there must necessarily exist a
certain classification of things according to their properties
—a classification which is either organically registered in
the system, as in the inferior creation, or is formed by
experience, as in ourselves.   And it may be further remarked, that the extent to which this classification is
carried, roughly indicates the height of intelligence—that,
while the lowest organisms are able to do little more than
discriminate organic from inorganic matter ; while the
generality of animals carry their classifications no further
than to a limited number of plants or creatures serving
for food, a limited number of beasts of prey, and a limited
number of places and materials ; the most degraded of the
human race possess a knowledge of the distinctive natures
of a great variety of substances, plants, animals, tools, persons, &c., not only as classes but as individuals.

What now is the mental process by which classification
is effected ?   Manifestly it is a recognition of the *likeness*
or *unlikeness* of things, either in respect of their sizes,
colours, forms, weights, textures, tastes, &c., or in respect
of their modes of action.   By some special mark, sound, or
motion, the savage identifies a certain four-legged creature he sees, as one that is good for food, and to be caught

in a particular way; or as one that is dangerous; and acts accordingly. He has classed together all the creatures that are *alike* in this particular. And manifestly in choosing the wood out of which to form his bow, the plant with which to poison his arrows, the bone from which to make his fish-hooks, he identifies them through their chief sensible properties as belonging to the general classes, wood, plant, and bone, but distinguishes them as belonging to sub-classes by virtue of certain properties in which they are *unlike* the rest of the general classes they belong to; and so forms genera and species.

And here it becomes manifest that not only is classification carried on by grouping together in the mind things that are *like*; but that classes and sub-classes are formed and arranged according to the *degrees of unlikeness*. Things widely contrasted are alone distinguished in the lower stages of mental evolution; as may be any day observed in an infant. And gradually as the powers of discrimination increase, the widely contrasted classes at first distinguished, come to be each divided into sub-classes, differing from each other less than the classes differ; and these sub-classes are again divided after the same manner. By the continuance of which process, things are gradually arranged into groups, the members of which are less and less *unlike*; ending, finally, in groups whose members differ only as individuals, and not specifically. And thus there tends ultimately to arise the notion of *complete likeness*. For manifestly, it is impossible that groups should continue to be sub-divided in virtue of smaller and smaller differences, without there being a simultaneous approximation to the notion of *no difference*.

Let us next notice that the recognition of likeness and unlikeness, which underlies classification, and out of which continued classification evolves the idea of complete likeness—let us next notice that it also underlies the process

of *naming*, and by consequence *language*. For all language consists, at the beginning, of symbols which are as
like to the things symbolized as it is practicable to make
them.   The language of signs is a means of conveying ideas
by mimicking the actions or peculiarities of the things referred to.   Verbal language is also, at the beginning, a
mode of suggesting objects or acts by imitating the sounds
which the objects make, or with which the acts are accompanied.   Originally these two languages were used simultaneously.   It needs but to watch the gesticulations with
which the savage accompanies his speech—to see a Bush-
man or a Kaffir dramatizing before an audience his mode
of catching game—or to note the extreme paucity of
words in all primitive vocabularies ; to infer that at first,
attitudes, gestures, and sounds, were all combined to pro·
duce as good a *likeness* as possible, of the things, animals,
persons, or events described ; and that as the sounds came
to be understood by themselves the gestures fell into dis-
use : leaving traces, however, in the manners of the more
excitable civilized races.   But be this as it may, it suffices
simply to observe, how many of the words current among
barbarous peoples are like the sounds appertaining to the
things signified ; how many of our own oldest and simplest
words have the same peculiarity ; how children tend to in-
vent imitative words ; and how the sign-language sponta-
neously formed by deaf mutes is invariably based upon
imitative actions—to at once see that the notion of *likeness*
is that from which the nomenclature of objects takes its
rise.

Were there space we might go on to point out how this
law of life is traceable, not only in the origin but in the development of language ; how in primitive tongues the plu-
ral is made by a duplication of the singular, which is a
multiplication of the word to make it *like* the multiplicity
of the things ; how the use of metaphor—that prolific

source of new words—is a suggesting of ideas that are *like* the ideas to be conveyed in some respect or other; and how, in the copious use of simile, fable, and allegory among uncivilized races, we see that complex conceptions, which there is yet no direct language for, are rendered, by presenting known conceptions more or less *like* them.

This view is further confirmed, and the predominance of this notion of likeness in primitive times further illustrated, by the fact that our system of presenting ideas to the eye originated after the same fashion. Writing and printing have descended from picture-language. The earliest mode of permanently registering a fact was by depicting it on a wall; that is—by exhibiting something as *like* to the thing to be remembered as it could be made. Gradually as the practice grew habitual and extensive, the most frequently repeated forms became fixed, and presently abbreviated; and, passing through the hieroglyphic and ideographic phases, the symbols lost all apparent relations to the things signified : just as the majority of our spoken words have done.

Observe again, that the same thing is true respecting the genesis of reasoning. The *likeness* that is perceived to exist between cases, is the essence of all early reasoning and of much of our present reasoning. The savage, having by experience discovered a relation between a certain object and a certain act, infers that the *like* relation will be found in future cases. And the expressions we constantly use in our arguments—" *analogy* implies," " the cases are not *parallel*," " by *parity* of reasoning," " there is no *similarity*,"—show how constantly the idea of likeness underlies our ratiocinative processes.

Still more clearly will this be seen on recognising the fact that there is a certain parallelism between reasoning and classification ; that the two have a common root; and that neither can go on without the other. For on the one hand, it is a familiar truth that the attributing to a body in consequence of some of its properties, all those other properties in virtue of which it is referred to a particular class, is an act of inference. And, on the other hand, the forming of a generalization is the putting together in one class, all those cases which present like relations ; while the drawing a deduction is essentially the perception that a particular case belongs to a certain class of cases previously generalized. So that as classification is a grouping together of *like things ;* reasoning is a grouping together of *like relations* among things. Add to which, that while the perfection gradually achieved in classification consists in the formation of groups of *objects* which are *completely alike ;* the perfection gradually achieved in reasoning consists in the formation of groups of *cases* which are *completely alike.*

Once more we may contemplate this dominant idea of likeness as exhibited in art. All art, civilized as well as savage, consists almost wholly in the making of objects *like* other objects ; either as found in Nature, or as produced by previous art. If we trace back the varied art-products now existing, we find that at each stage the divergence from previous patterns is but small when compared with the agreement ; and in the earliest art the persistency of imitation is yet more conspicuous. The old forms and ornaments and symbols were held sacred, and perpetually copied. Indeed, the strong imitative tendency notoriously displayed by the lowest human races, ensures among them a constant reproducing of likenesses of things, forms, signs, sounds, actions, and whatever else is imitable ; and we may even suspect that this aboriginal peculiarity is in some way connected with the culture and development of this general conception, which we have found so deep and wide-spread in its applications.

And now let us go on to consider how, by a further unfolding of this same fundamental notion, there is a gradual formation of the first germs of science. This idea of likeness which underlies classification, nomenclature, language spoken and written, reasoning, and art; and which plays so important a part because all acts of intelligence are made possible only by distinguishing among surrounding things, or grouping them into like and unlike;—this idea we shall find to be the one of which science is the especial product. Already during the stage we have been describing, there has existed *qualitative* prevision in respect to the commoner phenomena with which savage life is familiar; and we have now to inquire how the elements of *quantitative* prevision are evolved. We shall find that they originate by the perfecting of this same idea of likeness; that they have their rise in that conception of *complete likeness* which, as we have seen, necessarily results from the continued process of classification.

For when the process of classification has been carried as far as it is possible for the uncivilized to carry it—when the animal kingdom has been grouped not merely into quadrupeds, birds, fishes, and insects, but each of these divided into kinds—when there come to be sub-classes, in each of which the members differ only as individuals, and not specifically; it is clear that there must occur a frequent observation of objects which differ so little as to be indistinguishable. Among several creatures which the savage has killed and carried home, it must often happen that some one, which he wished to identify, is so exactly like another that he cannot tell which is which. Thus, then, there originates the notion of *equality*. The things which among ourselves are called *equal*—whether lines, angles, weights, temperatures, sounds or colours—are things which produce in us sensations that cannot be distinguished from each other. It is true that we now apply the word *equal* chiefly to the separate phenomena which objects exhibit, and not to groups of phenomena; but this limitation of the

9

idea has evidently arisen by subsequent analysis. And that the notion of equality did thus originate, will, we think, become obvious on remembering that as there were no artificial objects from which it could have been abstracted, it must have been abstracted from natural objects ; and that the various families of the animal kingdom chiefly furnish those natural objects which display the requisite exactitude of likeness.

The same order of experiences out of which this general idea of equality is evolved, gives birth at the same time to a more complex idea of equality ; or, rather, the process just described generates an idea of equality which further experience separates into two ideas—*equality of things* and *equality of relations.* While organic, and more especially animal forms, occasionally exhibit this perfection of likeness out of which the notion of simple equality arises, they more frequently exhibit only that kind of likeness which we call *similarity ;* and which is really compound equality. For the similarity of two creatures of the same species but of different sizes, is of the same nature as the similarity of two geometrical figures. In either case, any two parts of the one bear the same ratio to one another, as the homologous parts of the other. Given in any species, the proportions found to exist among the bones, and we may, and zoologists do, predict from any one, the dimensions of the rest ; just as, when knowing the proportions subsisting among the parts of a geometrical figure, we may, from the length of one, calculate the others. And if, in the case of similar geometrical figures, the similarity can be established only by proving exactness of proportion among the homologous parts ; if we express this relation between two parts in the one, and the corresponding parts in the other, by the formula A is to B as a is to b ; if we otherwise write this, A to B $= a$ to b ; if, consequently, the fact we prove is that the relation of A to B *equals* the relation of a to b ; then

it is manifest that the fundamental conception of similarity
is *equality of relations.*

With this explanation we shall be understood when we
say that the notion of equality of relations is the basis of
all exact reasoning. Already it has been shown that reasoning
in general is a recognition of *likeness* of relations; and
here we further find that while the notion of likeness of
things ultimately evolves the idea of simple equality, the
notion of likeness of relations evolves the idea of equality
of relations: of which the one is the concrete germ of ex·
act science, while the other is its abstract germ.

Those who cannot understand how the recognition of
similarity in creatures of the same kind, can have any alli-
ance with reasoning, will get over the difficulty on remembering that the phenomena among which equality of relations is thus perceived, are phenomena of the same order
and are present to the senses at the same time; while those
among which developed reason perceives relations, are generally neither of the same order, nor simultaneously present.
And if further, they will call to mind how Cuvier and Owen,
from a single part of a creature, as a tooth, construct the
rest by a process of reasoning based on this equality of relations, they will see that the two things are intimately
connected, remote as they at first seem. But we anticipate.
What it concerns us here to observe is, that from familiarity with organic forms there simultaneously arose the ideas
of *simple equality*, and *equality of relations.*

At the same time, too, and out of the same mental processes, came the first distinct ideas of *number.* In the earliest stages, the presentation of several like objects produced
merely an indefinite conception of multiplicity; as it still
does among Australians, and Bushmen, and Damaras, when
the number presented exceeds three or four. With such a
fact before us we may safely infer that the first clear numer·
ical conception was that of duality as contrasted with uni·

ty. And this notion of duality must necessarily have grown
up side by side with those of likeness and equality ; seeing
that it is impossible to recognise the likeness of two things
without also perceiving that there are two. From the
very beginning the conception of number must have been,
as it is still, associated with the likeness or equality of
the things numbered. If we analyze it, we find that simple enumeration is a registration of repeated impressions of any kind. That these may be capable of enu·
meration it is needful that they be more or less alike ; and
before any *absolutely true* numerical results can be reached, it is requisite that the units be *absolutely equal.* The
only way in which we can establish a numerical relationship between things that do not yield us like impressions,
is to divide them into parts that *do* yield us like impressions. Two unlike magnitudes of extension, force, time,
weight, or what not, can have their relative amounts estimated, only by means of some small unit that is contained
many times in both ; and even if we finally write down the
greater one as a unit and the other as a fraction of it, we
state, in the denominator of the fraction, the number of
parts into which the unit must be divided to be compara·
ble with the fraction.

It is, indeed, true, that by an evidently modern process of
abstraction, we occasionally apply numbers to unequal units,
as the furniture at a sale or the various animals on a farm,
simply as so many separate entities ; but no true result can
be brought out by calculation with units of this order.
And, indeed, it is the distinctive peculiarity of the calculus
in general, that it proceeds on the hypothesis of that abso·
lute equality of its abstract units, which no real units possess ; and that the exactness of its results holds only in
virtue of this hypothesis. The first ideas of number must
necessarily then have been derived from like or equal magnitudes as seen chiefly in organic objects ; and as the like

magnitudes most frequently observed were magnitudes of extension, it follows that geometry and arithmetic had a simultaneous origin.

Not only are the first distinct ideas of number co-ordin ate with ideas of likeness and equality, but the first efforts at numeration displayed the same relationship. On read- ing the accounts of various savage tribes, we find that the method of counting by the fingers, still followed by many children, is the aboriginal method. Neglecting the several cases in which the ability to enumerate does not reach even to the number of fingers on one hand, there are many cases in which it does not extend beyond ten—the limit of the simple finger notation. The fact that in so many instances, remote, and seemingly unrelated nations, have adopted *ten* as their basic number; together with the fact that in the re- maining instances the basic number is either *five* (the fingers of one hand) or *twenty* (the fingers and toes); almost of themselves show that the fingers were the original units of numeration. The still surviving use of the word *digit*, as the general name for a figure in arithmetic, is significant; and it is even said that our word *ten* (Sax. tyn; Dutch, tien; German, zehn) means in its primitive expanded form *two hands*. So that originally, to say there were ten things, was to say there were two hands of them.

From all which evidence it is tolerably clear that the earliest mode of conveying the idea of any number of things, was by holding up as many fingers as there were things; that is—using a symbol which was *equal*, in respect of multiplicity, to the group symbolized. For which infer- ence there is, indeed, strong confirmation in the recent statement that our own soldiers are even now spontaneous- ly adopting this device in their dealings with the Turks. And here it should be remarked that in this recombination of the notion of equality with that of multiplicity, by which the first steps in numeration are effected, we may see one of the earliest of those inosculations between the diverging branches of science, which are afterwards of perpetual occurrence.

Indeed, as this observation suggests, it will be well, before tracing the mode in which exact science finally emerges from the merely approximate judgments of the senses, and showing the non-serial evolution of its divisions, to note the non-serial character of those preliminary processes of which all after development is a continuation. On re-considering them it will be seen that not only are they divergent growths from a common root,—not only are they simultaneous in their progress; but that they are mutual aids; and that none can advance without the rest. That completeness of classification for which the unfolding of the perceptions paves the way, is impossible without a corresponding progress in language, by which greater varieties of objects are thinkable and expressible. On the one hand it is impossible to carry classification far without names by which to designate the classes; and on the other hand it is impossible to make language faster than things are classified.

Again, the multiplication of classes and the consequent narrowing of each class, itself involves a greater likeness among the things classed together; and the consequent approach towards the notion of complete likeness itself allows classification to be carried higher. Moreover, classification necessarily advances *pari passu* with rationality—the classification of *things* with the classification of *relations*. For things that belong to the same class are, by implication, things of which the properties and modes of behaviour—the co-existences and sequences—are more or less the same; and the recognition of this sameness of co-existences and sequences is reasoning. Whence it follows that the advance of classification is necessarily proportionate to the advance of generalizations. Yet further, the notion of *likeness*, both

in things and relations, simultaneously evolves by one process of culture the ideas of *equality* of things and *equality*
of relations; which are the respective bases of exact concrete reasoning and exact abstract reasoning—Mathematics
and Logic. And once more, this idea of equality, in the
very process of being formed, necessarily gives origin to
two series of relations—those of magnitude and those of
number: from which arise geometry and the calculus. Thus
the process throughout is one of perpetual subdivision and
perpetual intercommunication of the divisions. From the
very first there has been that *consensus* of different kinds of
knowledge, answering to the *consensus* of the intellectual
faculties, which, as already said, must exist among the sciences.

Let us now go on to observe how, out of the notions of
equality and *number*, as arrived at in the manner described,
there gradually arose the elements of quantitative prevision.

Equality, once having come to be definitely conceived,
was readily applicable to other phenomena than those of
magnitude. Being predicable of all things producing indistinguishable impressions, there naturally grew up ideas of
equality in weights, sounds, colours, &c.; and indeed it can
scarcely be doubted that the occasional experience of equal
weights, sounds, and colours, had a share in developing the
abstract conception of equality—that the ideas of equality
in size, relations, forces, resistances, and sensible properties in general, were evolved during the same period.
But however this may be, it is clear that as fast as the notion of equality gained definiteness, so fast did that lowest
kind of quantitative prevision which is achieved without
any instrumental aid, become possible.

The ability to estimate, however roughly, the amount
of a foreseen result, implies the conception that it will be
equal to a certain imagined quantity; and the correctness
of the estimate will manifestly depend upon the accuracy at

which the perceptions of sensible equality have arrived. A savage with a piece of stone in his hand, and another piece lying before him of greater bulk but of the same kind (a fact which he infers from the *equality* of the two in colour and texture) knows about what effort he must put forth to raise this other piece; and he judges accurately in proportion to the accuracy with which he perceives that the one is twice, three times, four times, &c. as large as the other; that is—in proportion to the precision of his ideas of equality and number. And here let us not omit to notice that even in these vaguest of quantitative previsions, the conception of *equality of relations* is also involved. For it is only in virtue of an undefined perception that the relation between bulk and weight in the one stone is *equal* to the relation between bulk and weight in the other, that even the roughest approximation can be made.

But how came the transition from those uncertain perceptions of equality which the unaided senses give, to the certain ones with which science deals? It came by placing the things compared in juxtaposition. Equality being predicated of things which give us indistinguishable impressions, and no accurate comparison of impressions being possible unless they occur in immediate succession, it results that exactness of equality is ascertainable in proportion to the closeness of the compared things. Hence the fact that when we wish to judge of two shades of colour whether they are alike or not, we place them side by side; hence the fact that we cannot, with any precision, say which of two allied sounds is the louder, or the higher in pitch, unless we hear the one immediately after the other; hence the fact that to estimate the ratio of weights, we take one in each hand, that we may compare their pressures by rapidly alternating in thought from the one to the other; hence the fact, that in a piece of music, we can continue to make equal beats when the first beat has been given, but cannot

ensure commencing with the same length of beat on a fu·
ture occasion; and hence, lastly, the fact, that of all magni·
tudes, those of *linear extension* are those of which the
equality is most accurately ascertainable, and those to
which by consequence all others have to be reduced. For
it is the peculiarity of linear extension that it alone allows
its magnitudes to be placed in *absolute* juxtaposition, or,
rather, in coincident position; it alone can test the equality
of two magnitudes by observing whether they will coalesce,
as two equal mathematical lines do, when placed between
the same points; it alone can test *equality* by trying wheth·
er it will become *identity*. Hence, then, the fact, that all
exact science is reducible, by an ultimate analysis, to results
measured in equal units of linear extension.

Still it remains to be noticed in what manner this deter·
mination of equality by comparison of linear magnitudes
originated. Once more may we perceive that surrounding
natural objects supplied the needful lessons. From the be·
ginning there must have been a constant experience of like
things placed side by side—men standing and walking to·
gether; animals from the same herd; fish from the same
shoal. And the ceaseless repetition of these experiences
could not fail to suggest the observation, that the nearer
together any objects were, the more visible became any in·
equality between them. Hence the obvious device of put·
ting in apposition, things of which it was desired to ascer·
tain the relative magnitudes. Hence the idea of *measure*.
And here we suddenly come upon a group of facts which
afford a solid basis to the remainder of our argument; while
they also furnish strong evidence in support of the forego·
ing speculations. Those who look sceptically on this at·
tempted rehabilitation of the earliest epochs of mental de·
velopment, and who more especially think that the derivation
of so many primary notions from organic forms is somewhat
strai ed, will perhaps see more probability in the several

hypotheses that have been ventured, on discovering that all measures of *extension* and *force* originated from the lengths and weights of organic bodies; and all measures of *time* from the periodic phenomena of either organic or inorganic bodies.

Thus, among linear measures, the cubit of the Hebrews was the *length of the forearm* from the elbow to the end of the middle finger; and the smaller scriptural dimensions are expressed in *hand-breadths* and *spans*. The Egyptian cubit, which was similarly derived, was divided into digits, which were *finger-breadths;* and each finger-breadth was more definitely expressed as being equal to four *grains of barley* placed breadthwise. Other ancient measures were the orgyia or *stretch of the arms*, the *pace*, and the *palm*. So persistent has been the use of these natural units of length in the East, that even now some of the Arabs mete out cloth by the forearm. So, too, is it with European measures. The *foot* prevails as a dimension throughout Europe, and has done since the time of the Romans, by whom, also, it was used: its lengths in different places varying not much more than men's feet vary. The heights of horses are still expressed in *hands*. The inch is the length of the terminal joint of *the thumb;* as is clearly shown in France, where *pouce* means both thumb and inch. Then we have the inch divided into three *barley-corns*.

So completely, indeed, have these organic dimensions served as the substrata of all mensuration, that it is only by means of them that we can form any estimate of some of the ancient distances. For example, the length of a degree on the Earth's surface, as determined by the Arabian astronomers shortly after the death of Haroun-al-Raschid, was fifty-six of their miles. We know nothing of their mile further than that it was 4000 cubits; and whether these were sacred cubits or common cubits, would remain doubtful, but that the length of the cubit is given as twenty-seven inches, and each inch defined as the thickness of six barley-grains. Thus one of the earliest measurements of a degree comes down to us in barley-grains. Not only did organic lengths furnish those approximate measures which satisfied men's needs in ruder ages, but they furnished also the standard measures required in later times. One instance occurs in our own history. To remedy the irregularities then prevailing, Henry I. commanded that the ulna, or ancient ell, which answers to the modern yard, should be made of the exact length of *his own arm.*

Measures of weight again had a like derivation. Seeds seem commonly to have supplied the unit. The original of the carat used for weighing in India is *a small bean.* Our own systems, both troy and avoirdupois, are derived primarily from wheat-corns. Our smallest weight, the grain, is *a grain of wheat.* This is not a speculation; it is an historically registered fact. Henry III. enacted that an ounce should be the weight of 640 dry grains of wheat from the middle of the ear. And as all the other weights are multiples or sub-multiples of this, it follows that the grain of wheat is the basis of our scale. So natural is it to use organic bodies as weights, before artificial weights have been established, or where they are not to be had, that in some of the remoter parts of Ireland the people are said to be in the habit, even now, of putting a man into the scales to serve as a measure for heavy commodities.

Similarly with time. Astronomical periodicity, and the periodicity of animal and vegetable life, are simultaneously used in the first stages of progress for estimating epochs. The simplest unit of time, the day, nature supplies ready made. The next simplest period, the mooneth or month, is also thrust upon men's notice by the conspicuous changes constituting a lunation. For larger divisions than these, the phenomena of the seasons, and the chief events from time to time occurring, have been used by early and uncivilized races. Among the Egyptians the rising of the Nile served as a mark. The New Zealanders were found to begin their year from the reappearance of the Pleiades above the sea. One of the uses ascribed to birds, by the Greeks, was to indicate the seasons by their migrations. Barrow describes the aboriginal Hottentot as denoting periods by the number of moons before or after the ripening of one of his chief articles of food. He further states that the Kaffir chronology is kept by the moon, and is registered by notches on sticks—the death of a favourite chief, or the gaining of a victory, serving for a new era. By which last fact, we are at once reminded that in early history, events are commonly recorded as occurring in certain reigns, and in certain years of certain reigns: a proceeding which practically made a king's reign a measure of duration.

And, as further illustrating the tendency to divide time by natural phenomena and natural events, it may be noticed that even by our own peasantry the definite divisions of months and years are but little used ; and that they habitually refer to occurrences as " before sheep-shearing," or " after harvest," or " about the time when the squire died." It is manifest, therefore, that the more or less equal periods perceived in Nature gave the first units of measure for time ; as did Nature's more or less equal lengths and weights give the first units of measure for space and force.

It remains only to observe, as further illustrating the evolution of quantitative ideas after this manner, that measures of value were similarly derived. Barter, in one form or other, is found among all but the very lowest human races. It is obviously based upon the notion of *equality of worth*. And as it gradually merges into trade

by the introduction of some kind of currency, we find
that the *measures of worth*, constituting this currency,
are organic bodies; in some cases *cowries*, in others
cocoa-nuts, in others *cattle*, in others *pigs ;* among the
American Indians peltry or *skins*, and in Iceland *dried
fish*.

Notions of exact equality and of measure having been
reached, there came to be definite ideas of relative magnitudes as being multiples one of another ; whence the practice of measurement by direct apposition of a measure.
The determination of linear extensions by this process can
scarcely be called science, though it is a step towards it;
but the determination of lengths of time by an analogous
process may be considered as one of the earliest samples of
quantitative prevision. For when it is first ascertained
that the moon completes the cycle of her changes in about
thirty days—a fact known to most uncivilized tribes that
can count beyond the number of their fingers—it is manifest that it becomes possible to say in what number of days
any specified phase of the moon will recur ; and it is also
manifest that this prevision is effected by an opposition of
two times, after the same manner that linear space is measured by the opposition of two lines. For to express the
moon's period in days, is to say how many of these units
of measure are contained in the period to be measured—is
to ascertain the distance between two points in time by
means of a *scale of days*, just as we ascertain the distance
between two points in space by a scale of feet or inches :
and in each case the scale coincides with the thing measured—mentally in the one ; visibly in the other. So that
in this simplest, and perhaps earliest case of quantitative
prevision, the phenomena are not only thrust daily upon
men's notice, but Nature is, as it were, perpetually repeating that process of measurement by observing which
the prevision is effected. And thus there may be signi-

ficance in the remark which some have made, that alike in Hebrew, Greek, and Latin, there is an affinity be-tween the word meaning moon, and that meaning measure.

This fact, that in very early stages of social progress it is known that the moon goes through her changes in about thirty days, and that in about twelve moons the seasons return—this fact that chronological astronomy assumes a certain scientific character even before geometry does; while it is partly due to the circumstance that the astro-nomical divisions, day, month, and year, are ready made for us, is partly due to the further circumstances that agricultural and other operations were at first regulated astronomically, and that from the supposed divine nature of the heavenly bodies their motions determined the periodical religious festivals. As instances of the one we have the observation of the Egyptians, that the rising of the Nile corresponded with the heliacal rising of Sirius; the directions given by Hesiod for reaping and ploughing, according to the positions of the Pleiades; and his maxim that "fifty days after the turning of the sun is a seasonable time for beginning a voyage." As instances of the other, we have the naming of the days after the sun, moon, and planets; the early attempts among Eastern nations to regulate the calendar so that the gods might not be offend-ed by the displacement of their sacrifices; and the fix-ing of the great annual festival of the Peruvians by the position of the sun. In all which facts we see that, at first, science was simply an appliance of religion and industry.

After the discoveries that a lunation occupies nearly thirty days, and that some twelve lunations occupy a year —discoveries of which there is no historical account, but which may be inferred as the earliest, from the fact that existing uncivilized races have made them—we come to the first known astronomical records, which are those of eclipses. The Chaldeans were able to predict these. "This they did, probably," says Dr. Whewell in his useful history, from which most of the materials we are about to use will be drawn, " by means of their cycle of 223 months, or about eighteen years; for at the end of this time, the eclipses of the moon begin to return, at the same intervals and in the same order as at the beginning." Now this method of calculating eclipses by means of a recurring cycle,— the *Saros* as they called it—is a more complex case of prevision by means of coincidence of measures. For by what observations must the Chaldeans have discovered this cycle? Obviously, as Delambre infers, by inspecting their registers; by comparing the successive intervals; by finding that some of the intervals were alike; by seeing that these equal intervals were eighteen years apart; by discovering that *all* the intervals that were eighteen years apart were equal; by ascertaining that the intervals formed a series which repeated itself, so that if one of the cycles of intervals were superposed on another the divisions would fit. This once perceived, and it manifestly became possible to use the cycle as a scale of time by which to measure out future periods. Seeing thus that the process of so predicting eclipses, is in essence the same as that of predicting the moon's monthly changes by observing the number of days after which they repeat—seeing that the two differ only in the extent and irregularity of the intervals, it is not difficult to understand how such an amount of knowledge should so early have been reached. And we shall be less surprised, on remembering that the only things involved in these previsions were *time* and *number;* and that the time was in a manner self-numbered.

Still, the ability to predict events recurring only after so long a period as eighteen years, implies a considerable advance in civilization—a considerable development of general knowledge; and we have now to inquire what progress in other sciences accompanied, and was necessary to, these astronomical previsions. In the first place, there must clearly have been a tolerably efficient system of calculation. Mere finger-counting, mere head-reckoning, even with the aid of a regular decimal notation, could not have sufficed for numbering the days in a year; much less the years, months, and days between eclipses. Consequently there must have been a mode of registering numbers; probably even a system of numerals. The earliest numerical records, if we may judge by the practices of the less civilized races now existing, were probably kept by notches cut on sticks, or strokes marked on walls; much as public-house scores are kept now. And there seems reason to believe that the first numerals used were simply groups of straight strokes, as some of the still-extant Roman ones are; leading us to suspect that these groups of strokes were used to represent groups of fingers, as the groups of fingers had been used to represent groups of objects—a supposition quite in conformity with the aboriginal system of picture writing and its subsequent modifications. Be this so or not, however, it is manifest that before the Chaldeans discovered their *Saros*, there must have been both a set of written symbols serving for an extensive numeration, and a familiarity with the simpler rules of arithmetic.

Not only must abstract mathematics have made some progress, but concrete mathematics also. It is scarcely possible that the buildings belonging to this era should have been laid out and erected without any knowledge of geometry. At any rate, there must have existed that elementary geometry which deals with direct measurement—with the apposition of lines; and it seems that only after the discovery of those simple proceedings, by which right angles are drawn, and relative positions fixed, could so regular an architecture be executed. In the case of the other division of concrete mathematics—mechanics, we have definite evidence of progress. We know that the lever and the inclined plane were employed during this period: implying that there was a qualitative prevision of their effects, though not a quantitative one. But we know more. We read of weights in the earliest records; and we find weights in ruins of the highest antiquity. Weights imply scales, of which we have also mention; and scales involve the primary theorem of mechanics in its least complicated form —involve not a qualitative but a quantitative prevision of mechanical effects. And here we may notice how mechanics, in common with the other exact sciences, took its rise from the simplest application of the idea of *equality*. For the mechanical proposition which the scales involve, is, that if a lever with *equal* arms, have *equal* weights suspended from them, the weights will remain at *equal* altitudes. And we may further notice, how, in this first step of rational mechanics, we see illustrated that truth awhile since referred to, that as magnitudes of linear extension are the only ones of which the equality is exactly ascertainable, the equalities of other magnitudes have at the outset to be determined by means of them. For the equality of the weights which balance each other in scales, wholly depends upon the equality of the arms: we can know that the weights are equal only by proving that the arms are equal. And when by this means we have obtained a system of weights,—a set of equal units of force, then does a science of mechanics become possible. Whence, indeed, it follows, that rational mechanics could not possibly have any other starting-point than the scales.

Let us further remember, that during this same period there was a limited knowledge of chemistry. The many arts which we know to have been carried on must have been impossible without a generalized experience of the modes in which certain bodies affect each other under special conditions. In metallurgy, which was extensively practised, this is abundantly illustrated. And we even have evidence that in some cases the knowledge possessed was, in a sense, quantitative. For, as we find by analysis that the hard alloy of which the Egyptians made their cutting tools, was composed of copper and tin in fixed proportions, there must have been an established prevision that such an alloy was to be obtained only by mixing them in these proportions. It is true, this was but a simple empirical generalization; but so was the generalization respecting the recurrence of eclipses; so are the first generalizations of every science.

Respecting the simultaneous advance of the sciences during this early epoch, it only remains to remark that even the most complex of them must have made some progress—perhaps even a greater relative progress than any of the rest. For under what conditions only were the foregoing developments possible? There first required an established and organized social system. A long continued registry of eclipses; the building of palaces; the use of scales; the practice of metallurgy—alike imply a fixed and populous nation. The existence of such a nation not only presupposes laws, and some administration of justice, which we know existed, but it presupposes successful laws—laws conforming in some degree to the conditions of social stability—laws enacted because it was seen that the actions forbidden by them were dangerous to the State. We do not by any means say that all, or even the greater part, of the laws were of this nature; but we do say, that the fundamental ones were. It cannot be denied that the laws affecting life and property were such. It cannot be denied that, however little these were enforced between class and class, they were to a considerable extent enforced between members of the same class. It can scarcely be questioned, that the administration of them between members of the same class was seen by rulers to be necessary for keeping their subjects together. And knowing, as we do, that, other things equal, nations prosper in proportion to the justness of their arrangements, we may fairly infer that the very cause of the advance of these earliest nations out of aboriginal barbarism, was the greater recognition among them of the claims to life and property.

But supposition aside, it is clear that the habitual recognition of these claims in their laws, implied some prevision of social phenomena. Even thus early there was a certain amount of social science. Nay, it may even be shown that there was a vague recognition of that fundamental principle on which all the true social science is based—the equal rights of all to the free exercise of their faculties. That same idea of *equality*, which, as we have seen, underlies all other science, underlies also morals and sociology. The conception of justice, which is the primary one in morals; and the administration of justice, which is the vital condition of social existence; are impossible, without the recognition of a certain likeness in men's claims, in virtue of their common humanity. *Equity* literally means *equalness;* and if it be admitted that there were even the vaguest ideas of equity in these primitive eras, it must be admitted that there was some appreciation of the equalness of men's liberties to pursue the objects of life—some appreciation, therefore, of the essential principle of national equilibrium.

Thus in this initial stage of the positive sciences, before geometry had yet done more than evolve a few empirical rules—before mechanics had passed beyond its first theorem—before astronomy had advanced from its merely chronological phase into the geometrical; the most involved of the sciences had reached a certain degree of development —a development without which no progress in other sciences was possible.

Only noting as we pass, how, thus early, we may see that the progress of exact science was not only towards an

increasing number of previsions, but towards previsions
more accurately quantitative—how, in astronomy, the recurring period of the moon's motions was by and by more
correctly ascertained to be nineteen years, or two hundred
and thirty-five lunations; how Callipus further corrected
this Metonic cycle, by leaving out a day at the end of every
seventy-six years; and how these successive advances im
plied a longer continued registry of observations, and the
co-ordination of a greater number of facts—let us go on to
inquire how geometrical astronomy took its rise.

The first astronomical instrument was the gnomon
This was not only early in use in the East, but it was found
also among the Mexicans; the sole astronomical observations of the Peruvians were made by it; and we read that
1100 B.C., the Chinese found that, at a certain place, the
length of the sun's shadow, at the summer solstice, was to
the height of the gnomon, as one and a half to eight.
Here again it is observable, not only that the instrument is
found ready made, but that Nature is perpetually perform-
ing the process of measurement. Any fixed, erect object
—a column, a dead palm, a pole, the angle of a building—
serves for a gnomon; and it needs but to notice the changing position of the shadow it daily throws, to make the
first step in geometrical astronomy. How small this first
step was, may be seen in the fact that the only things ascertained at the outset were the periods of the summer
and winter solstices, which corresponded with the least and
greatest lengths of the mid-day shadow; and to fix which,
it was needful merely to mark the point to which each
day's shadow reached.

And now let it not be overlooked that in the observing
at what time during the next year this extreme limit of the
shadow was again reached, and in the inference that the
sun had then arrived at the same turning point in his annual course, we have one of the simplest instances of that

combined use of *equal magnitudes* and *equal relations*, by
which all exact science, all quantitative prevision, is reached.
For the relation observed was between the length of the
sun's shadow and his position in the heavens; and the inference drawn was that when, next year, the extremity of
his shadow came to the same point, he occupied the same
place. That is, the ideas involved were, the equality of the
shadows, and the equality of the relations between shadow
and sun in successive years. As in the case of the scales,
the equality of relations here recognized is of the simplest
order. It is not as those habitually dealt with in the higher
kinds of scientific reasoning, which answer to the general
type—the relation between two and three equals the relation between six and nine; but it follows the type—the relation between two and three, equals the relation between
two and three; it is a case of not simply *equal* relations,
but *coinciding* relations. And here, indeed, we may see
beautifully illustrated how the idea of equal relations takes
its rise after the same manner that that of equal magnitude
does. As already shown, the idea of equal magnitudes
arose from the observed coincidence of two lengths placed
together; and in this case we have not only two coincident
lengths of shadows, but two coincident relations between
sun and shadows.

From the use of the gnomon there naturally grew up
the conception of angular measurements; and with the
advance of geometrical conceptions there came the hemi-
sphere of Berosus, the equinoctial armil, the solstitial armil,
and the quadrant of Ptolemy—all of them employing shadows as indices of the sun's position, but in combination
with angular divisions. It is obviously out of the question
for us here to trace these details of progress. It must suffice to remark that in all of them we may see that notion
of equality of relations of a more complex kind, which is
best illustrated in the astrolabe, an instrument which con-

sisted " of circular rims, moveable one within the other, or
about poles, and contained circles which were to be brought
into the position of the ecliptic, and of a plane passing
through the sun and the poles of the ecliptic "—an instrument, therefore, which represented, as by a model, the relative positions of certain imaginary lines and planes in the
heavens; which was adjusted by putting these representative lines and planes into parallelism and coincidence with
the celestial ones; and which depended for its use upon the
perception that the relations between these representative
lines and planes were *equal* to the relations between those
represented.

Were there space, we might go on to point out how the
conception of the heavens as a revolving hollow sphere,
the discovery of the globular form of the earth, the explanation of the moon's phases, and indeed all the successive
steps taken, involved this same mental process. But we
must content ourselves with referring to the theory of eccentrics and epicycles, as a further marked illustration of
it. As first suggested, and as proved by Hipparchus to afford an explanation of the leading irregularities in the celestial motions, this theory involved the perception that
the progressions, retrogressions, and variations of velocity
seen in the heavenly bodies, might be reconciled with their
assumed uniform movement in circles, by supposing that
the earth was not in the centre of their orbits; or by supposing that they revolved in circles whose centres revolved
round the earth; or by both. The discovery that this
would account for the appearances, was the discovery that
in certain geometrical diagrams the relations were such,
that the uniform motion of a point would, when looked at
from a particular position, present analogous irregularities;
and the calculations of Hipparchus involved the belief that the
relations subsisting among these geometrical curves were
equal to the relations subsisting among the celestial orbits.

Leaving here these details of astronomical progress, and the philosophy of it, let us observe how the relatively concrete science of geometrical astronomy, having been thus far helped forward by the development of geometry in general, reacted upon geometry, caused it also to advance, and was again assisted by it. Hipparchus, before making his solar and lunar tables, had to discover rules for calculating the relations between the sides and angles of triangles—*trigonometry* a subdivision of pure mathematics. Further, the reduction of the doctrine of the sphere to the quantitative form needed for astronomical purposes, required the formation of a *spherical trigonometry,* which was also achieved by Hipparchus. Thus both plane and spherical trigonometry, which are parts of the highly abstract and simple science of extension, remained undeveloped until the less abstract and more complex science of the celestial motions had need of them. The fact admitted by M. Comte, that since Descartes the progress of the abstract division of mathematics has been determined by that of the concrete division, is paralleled by the still more significant fact that even thus early the progress of mathematics was determined by that of astronomy.

And here, indeed, we may see exemplified the truth, which the subsequent history of science frequently illustrates, that before any more abstract division makes a further advance, some more concrete division must suggest the necessity for that advance—must present the new order of questions to be solved. Before astronomy presented Hipparchus with the problem of solar tables, there was nothing to raise the question of the relations between lines and angles; the subject-matter of trigonometry had not been conceived. And as there must be subject-matter before there can be investigation, it follows that the progress of the concrete divisions is as necessary to that of the abstract, as the progress of the abstract to that of the concrete.

Just incidentally noticing the circumstance that the
epoch we are describing witnessed the evolution of algebra,
a comparatively abstract division of mathematics, by the
union of its less abstract divisions, geometry and arithmetic—a fact proved by the earliest extant samples of algebra, which are half algebraic, half geometric—we go on to
observe that during the era in which mathematics and
astronomy were thus advancing, rational mechanics made
its second step; and something was done towards giving a
quantitative form to hydrostatics, optics, and harmonics.
In each case we shall see as before, how the idea of equality underlies all quantitative prevision; and in what simple
forms this idea is first applied.

As already shown, the first theorem established in mechanics was, that equal weights suspended from a lever with
equal arms would remain in equilibrium. Archimedes discovered that a lever with unequal arms was in equilibrium
when one weight was to its arm as the other arm to its
weight; that is—when the numerical relation between one
weight and its arm was *equal* to the numerical relation between the other arm and its weight.

The first advance made in hydrostatics, which we also
owe to Archimedes, was the discovery that fluids press
equally in all directions; and from this followed the solution of the problem of floating bodies: namely, that they
are in equilibrium when the upward and downward pressures are *equal*.

In optics, again, the Greeks found that the angle of incidence is *equal* to the angle of reflection; and their knowledge reached no further than to such simple deductions
from this as their geometry sufficed for. In harmonics
they ascertained the fact that three strings of *equal* lengths
would yield the octave, fifth and fourth, when strained by
weights having certain definite ratios; and they did not
progress much beyond this. In the one of which cases we

see geometry used in elucidation of the laws of light; and in the other, geometry and arithmetic made to measure the phenomena of sound.

Did space permit, it would be desirable here to describe the state of the less advanced sciences—to point out how, while a few had thus reached the first stages of quantitative prevision, the rest were progressing in qualitative prevision—how some small generalizations were made respecting evaporation, and heat, and electricity, and magnetism, which, empirical as they were, did not in that respect differ from the first generalizations of every science—how the Greek physicians had made advances in physiology and pathology, which, considering the great imperfection of our present knowledge, are by no means to be despised—how zoology had been so far systematized by Aristotle, as, to some extent, enabled him from the presence of certain organs to predict the presence of others—how in Aristotle's *Politics*, there is some progress towards a scientific conception of social phenomena, and sundry previsions respecting them—and how in the state of the Greek societies, as well as in the writings of Greek philosophers, we may recognise not only an increasing clearness in that conception of equity on which the social science is based, but also some appreciation of the fact that social stability depends upon the maintenance of equitable regulations. We might dwell at length upon the causes which retarded the development of some of the sciences, as for example, chemistry: showing that relative complexity had nothing to do with it—that the oxidation of a piece of iron is a simpler phenomenon than the recurrence of eclipses, and the discovery of carbonic acid less difficult than that of the precession of the equinoxes—but that the relatively slow advance of chemical knowledge was due, partly to the fact that its phenomena were not daily thrust on men's notice as those of astronomy were; partly to the fact that Nature

10

does not habitually supply the means, and suggest the
modes of investigation, as in the sciences dealing with time
extension, and force ; and partly to the fact that the great
majority of the materials with which chemistry deals, in
stead of being ready to hand, are made known only by the
arts in their slow growth ; and partly to the fact that even
when known, their chemical properties are not self-exhibit-
ed, but have to be sought out by experiment.

Merely indicating all these considerations, however, let
us go on to contemplate the progress and mutual influence
of the sciences in modern days; only parenthetically noticing how, on the revival of the scientific spirit, the successive stages achieved exhibit the dominance of the same
law hitherto traced—how the primary idea in dynamics, a
uniform force, was defined by Galileo to be a force which
generates *equal* velocities in *equal* successive times—how
the uniform action of gravity was first experimentally determined by showing that the time elapsing before a body
thrown up, stopped, was *equal* to the time it took to fall—
how the first fact in compound motion which Galileo ascertained was, that a body projected horizontally will have a
uniform motion onwards and a uniformly accelerated motion downwards ; that is, will describe *equal* horizontal
spaces in *equal* times, compounded with *equal* vertical in-
crements in *equal* times—how his discovery respecting the
pendulum was, that its oscillations occupy *equal* intervals
of time whatever their length—how the principle of virtual
velocities which he established is, that in any machine the
weights that balance each other, are reciprocally as their
virtual velocities ; that is, the relation of one set of weights
to their velocities *equals* the relation of the other set of
velocities to their weights ;—and how thus his achievements consisted in showing the equalities of certain magnitudes and relations, whose equalities had not been previously recognised.

When mechanics had reached the point to which Galileo brought it—when the simple laws of force had been disentangled from the friction and atmospheric resistance by which all their earthly manifestations are disguised—when progressing knowledge of *physics* had given a due insight into these disturbing causes—when, by an effort of abstraction, it was perceived that all motion would be uniform and rectilinear unless interfered with by external forces—and when the various consequences of this perception had been worked out; then it became possible, by the union of geometry and mechanics, to initiate physical astronomy. Geometry and mechanics having diverged from a common root in men's sensible experiences; having, with occasional inosculations, been separately developed, the one partly in connexion with astronomy, the other solely by analyzing terrestrial movements; now join in the investigations of Newton to create a true theory of the celestial motions. And here, also, we have to notice the important fact that, in the very process of being brought jointly to bear upon astronomical problems, they are themselves raised to a higher phase of development. For it was in dealing with the questions raised by celestial dynamics that the then incipient infinitesimal calculus was unfolded by Newton and his continental successors; and it was from inquiries into the mechanics of the solar system that the general theorems of mechanics contained in the " Principia,"—many of them of purely terrestrial application—took their rise. Thus, as in the case of Hipparchus, the presentation of a new order of concrete facts to be analyzed, led to the discovery of new abstract facts; and these abstract facts having been laid hold of, gave means of access to endless groups of concrete facts before incapable of quantitative treatment.

Meanwhile, physics had been carrying further that progress without which, as just shown, rational mechanics could not be disentangled. In hydrostatics, Stevinus had extended and applied the discovery of Archimedes. Torricelli had proved atmospheric pressure, "by showing that this pressure sustained different liquids at heights inversely proportional to their densities;" and Pascal "established the necessary diminution of this pressure at increasing heights in the atmosphere:" discoveries which in part reduced this branch of science to a quantitative form. Something had been done by Daniel Bernouilli towards the dynamics of fluids. The thermometer had been invented; and a number of small generalizations reached by it. Huyghens and Newton had made considerable progress in optics; Newton had approximately calculated the rate of transmission of sound; and the continental mathematicians had succeeded in determining some of the laws of sonorous vibrations. Magnetism and electricity had been considerably advanced by Gilbert. Chemistry had got as far as the mutual neutralization of acids and alkalies. And Leonardo da Vinci had advanced in geology to the conception of the deposition of marine strata as the origin of fossils. Our present purpose does not require that we should give particulars. All that it here concerns us to do is to illustrate the *consensus* subsisting in this stage of growth, and afterwards. Let as look at a few cases.

The theoretic law of the velocity of sound enunciated by Newton on purely mechanical considerations, was found wrong by one-sixth. The error remained unaccounted for until the time of Laplace, who, suspecting that the heat disengaged by the compression of the undulating strata of the air, gave additional elasticity, and so produced the difference, made the needful calculations and found he was ight. Thus acoustics was arrested until thermology overtook and aided it. When Boyle and Marriot had discovered the relation between the density of gases and the pressures they are subject to; and when it thus became possible to calculate the rate of decreasing density in the upper parts of the atmosphere; it also became possible to make approximate tables of the atmospheric refraction of light. Thus optics, and with it astronomy, advanced with barology. After the discovery of atmospheric pressure had led to the invention of the air-pump by Otto Guericke; and after it had become known that evaporation increases in rapidity as atmospheric pressure decreases; it became possible for Leslie, by evaporation in a vacuum, to produce the greatest cold known; and so to extend our knowledge of thermology by showing that there is no zero within reach of our researches. When Fourier had determined the laws of conduction of heat, and when the Earth's temperature had been found to increase below the surface one degree in every forty yards, there were data for inferring the past condition of our globe; the vast period it has taken to cool down to its present state; and the immense age of the solar system—a purely astronomical consideration.

Chemistry having advanced sufficiently to supply the needful materials, and a physiological experiment having furnished the requisite hint, there came the discovery of galvanic electricity. Galvanism reacting on chemistry disclosed the metallic bases of the alkalies, and inaugurated the electro-chemical theory; in the hands of Oersted and Ampère it led to the laws of magnetic action; and by its aid Faraday has detected significant facts relative to the constitution of light. Brewster's discoveries respecting double refraction and dipolarization proved the essential truth of the classification of crystalline forms according to the number of axes, by showing that the molecular constitution depends upon the axes. In these and in numerous other cases, the mutual influence of the sciences has been quite independent of any supposed hierarchical order. Often, too, their inter-actions are more complex than as

thus instanced—involve more sciences than two.   One
illustration of this must suffice.   We quote it in full from
the *History of the Inductive Sciences.*   In Book XI., chap.
II., on "The Progress of the Electrical Theory," Dr
Whewell writes :—

"Thus at that period, mathematics was behind experiment,
and a problem was proposed, in which theoretical results were
wanted for comparison with observation, but could not be accurately obtained; as was the case in astronomy also, till the time
of the approximate solution of the problem of three bodies, and
the consequent formation of the tables of the moon and planets,
on the theory of universal gravitation.   After some time, electrical theory was relieved from this reproach, mainly in consequence of the progress which astronomy had occasioned in pure
mathematics.   About 1801 there appeared in the *Bulletin des
Sciences,* an exact solution of the problem of the distribution of
electric fluid on a spheroid, obtained by Biot, by the application
of the peculiar methods which Laplace had invented for the problem of the figure of the planets.   And, in 1811, M. Poisson applied
Laplace's artifices to the case of two spheres acting upon one
another in contact, a case to which many of Coulomb's experiments were referrible; and the agreement of the results of
theory and observation, thus extricated from Coulomb's numbers obtained above forty years previously, was very striking and
convincing."

Not only do the sciences affect each other after this
direct manner, but they affect each other indirectly.
Where there is no dependence, there is yet analogy—
equality of relations; and the discovery of the relations
subsisting among one set of phenomena, constantly suggests a search for the same relations among another set.
Thus the established fact that the force of gravitation varies
inversely as the square of the distance, being recognized as
a necessary characteristic of all influences proceeding from
a centre, raised the suspicion that heat and light follow the
same law ; which proved to be the case—a suspicion and a

confirmation which were repeated in respect to the electric and magnetic forces. Thus again the discovery of the polarization of light led to experiments which ended in the discovery of the polarization of heat—a discovery that could never have been made without the antecedent one. Thus, too, the known refrangibility of light and heat lately produced the inquiry whether sound also is not refrangible; which on trial it turns out to be.

In some cases, indeed, it is only by the aid of conceptions derived from one class of phenomena that hypotheses respecting other classes can be formed. The theory, at one time favoured, that evaporation is a solution of water in air, was an assumption that the relation between water and air is *like* the relation between salt and water; and could never have been conceived if the relation between salt and water had not been previously known. Similarly the received theory of evaporation—that it is a diffusion of the particles of the evaporating fluid in virtue of their atomic repulsion—could not have been entertained without a foregoing experience of magnetic and electric repulsions. So complete in recent days has become this *consensus* among the sciences, caused either by the natural entanglement of their phenomena, or by analogies in the relations of their phenomena, that scarcely any considerable discovery concerning one order of facts now takes place, without very shortly leading to discoveries concerning other orders.

To produce a tolerably complete conception of this process of scientific evolution, it would be needful to go back to the beginning, and trace in detail the growth of classifications and nomenclatures; and to show how, as subsidiary to science, they have acted upon it, and it has reacted upon them. We can only now remark that, on the one hand, classifications and nomenclatures have aided science by continually subdividing the subject-matter of research, and giving fixity and diffusion to the truths disclosed; and that on the other hand, they have caught from it that increasing quantitativeness, and that progress from considerations touching single phenomena to considerations touching the relations among many phenomena, which we have been describing.

Of this last influence a few illustrations must be given. In chemistry it is seen in the facts, that the dividing of matter into the four elements was ostensibly based upon the single property of weight; that the first truly chemical division into acid and alkaline bodies, grouped together bodies which had not simply one property in common, but in which one property was constantly related to many others; and that the classification now current, places together in groups *supporters of combustion, metallic and non-metallic bases, acids, salts,* &c., bodies which are often quite unlike in sensible qualities, but which are like in the majority of their *relations* to other bodies. In mineralogy again, the first classifications were based upon differences in aspect, texture, and other physical attributes. Berzelius made two attempts at a classification based solely on chemical constitution. That now current, recognises as far as possible the *relations* between physical and chemical characters. In botany the earliest classes formed were *trees, shrubs,* and *herbs:* magnitude being the basis of distinction. Dioscorides divided vegetables into *aromatic, alimentary, medicinal,* and *vinous:* a division of chemical character. Cæsalpinus classified them by the seeds, and seed-vessels, which he preferred because of the *relations* found to subsist between the character of the fructification and the general character of the other parts.

While the "natural system" since developed, carrying out the doctrine of Linnæus, that "natural orders must be formed by attention not to one or two, but to *all* the parts of plants," bases its divisions on like peculiarities which are found to be *constantly related* to the greatest number of other like peculiarities.   And similarly in zoology, the successive classifications, from having been originally determined by external and often subordinate characters not indicative of the essential nature, have been gradually more and more determined by those internal and fundamental differences, which have uniform *relations* to the greatest number of other differences.   Nor shall we be surprised at this analogy between the modes of progress of positive science and classification, when we bear in mind that both proceed by making generalizations; that both enable us to make previsions differing only in their precision; and that while the one deals with equal properties and relations, the other deals with properties and relations that approximate towards equality in variable degrees.

Without further argument, it will, we think, be sufficiently clear that the sciences are none of them separately evolved—are none of them independent either logically or historically; but that all of them have, in a greater or less degree, required aid and reciprocated it.   Indeed, it needs but to throw aside theses, and contemplate the mixed character of surrounding phenomena, to at once see that these notions of division and succession in the kinds of knowledge are none of them actually true, but are simple scientific fictions . good, if regarded merely as aids to study; bad, if regarded as representing realities in Nature.   Consider them critically, and no facts whatever are presented to our senses uncombined with other facts—no facts whatever but are in some degree disguised by accompanying facts: disguised in such a manner that all must be partially understood before any one can be understood.   If it be said, as by M. Comte, that gravitating force should be treated of before other forces, seeing that all things are subject to it, it may on like grounds be said that heat should be first dealt with; seeing that thermal forces are everywhere in

action ; that the ability of any portion of matter to mani
fest visible gravitative phenomena depends on its state of
aggregation, which is determined by heat; that only by
the aid of thermology can we explain those apparent exceptions to the gravitating tendency which are presented
by steam and smoke, and so establish its universality, and
that, indeed, the very existence of the solar system in a solid form is just as much a question of heat as it is one of
gravitation.

Take other cases :—All phenomena recognised by the
eyes, through which only are the data of exact science ascertainable, are complicated with optical phenomena ; and
cannot be exhaustively known until optical principles are
known. The burning of a candle cannot be explained
without involving chemistry, mechanics, thermology.
Every wind that blows is determined by influences partly
solar, partly lunar, partly hygrometric ; and implies considerations of fluid equilibrium and physical geography
The direction, dip, and variations of the magnetic needle,
are facts half terrestrial, half celestial—are caused by earthly forces which have cycles of change corresponding with
astronomical periods. The flowing of the gulf-stream and the
annual migration of icebergs towards the equator, depending as they do on the balancing of the centripetal and centrifugal forces acting on the ocean, involve in their explanation the Earth's rotation and spheroidal form, the laws of
hydrostatics, the relative densities of cold and warm water,
and the doctrines of evaporation. It is no doubt true, as
M. Comte says, that " our position in the solar system, and
the motions, form, size, equilibrium of the mass of our
world among the planets, must be known before we can understand the phenomena going on at its surface." But, fa-
tally for his hypothesis, it is also true that we must understand a great part of the phenomena going on at its surface
before we can know its position, &c., in the solar system

It is not simply that, as we have already shown, those geometrical and mechanical principles by which celestial appearances are explained, were first generalized from terrestrial experiences ; but it is that the very obtainment of correct data, on which to base astronomical generalizations,
implies advanced terrestrial physics.

Until after optics had made considerable advance, the
Copernican system remained but a speculation. A single
modern observation on a star has to undergo a careful anal·
ysis by the combined aid of various sciences—has to *be digest-
ed by the organism of the sciences ;* which have severally
to assimilate their respective parts of the observation, before the essential fact it contains is available for the further
development of astronomy.   It has to be corrected not
only for nutation of the earth's axis and for precession of
the equinoxes, but for aberration and for refraction ; and
the formation of the tables by which refraction is calculated, presupposes knowledge of the law of decreasing density
in the upper atmospheric strata ; of the law of decreasing
temperature, and the influence of this on the density ; and of
hygrometric laws as also affecting density.   So that, to get
materials for further advance, astronomy requires not only
the indirect aid of the sciences which have presided over
the making of its improved instruments, but the direct aid
of an advanced optics, of barology, of thermology, of hy·
grometry ; and if we remember that these delicate obser·
vations are in some cases registered electrically, and that
they are further corrected for the " personal equation "—the
time elapsing between seeing and registering, which varies
with different observers—we may even add electricity and
psychology. If, then, so apparently simple a thing as ascertaining the position of a star is complicated with so
many phenomena, it is clear that this notion of the independence of the sciences, or certain of them, will not hold.

Whether objectively independent or not, they cannot

be subjectively so—they cannot have independence as presented to our consciousness; and this is the only kind of
independence with which we are concerned.   And here,
before leaving these illustrations, and especially this last
one, let us not omit to notice how clearly they exhibit that
increasingly active *consensus* of the sciences which characterizes their advancing development.   Besides finding that
in these later times a discovery in one science commonly
causes progress in others ; besides finding that a great part
of the questions with which modern science deals are so mixed as to require the co-operation of many sciences for their
solution ; we find in this last case that, to make a single good
observation in the purest of the natural sciences, requires
the combined assistance of half a dozen other sciences.

Perhaps the clearest comprehension of the interconnect-
ed growth of the sciences may be obtained by contemplating that of the arts, to which it is strictly analogous, and
with which it is inseparably bound up.   Most intelligent
persons must have been, at one time or other, struck with
the vast array of antecedents pre-supposed by one of our
processes of manufacture.   Let him trace the production
of a printed cotton, and consider all that is implied by it.
There are the many successive improvements through
which the power-looms reached their present perfection;
there is the steam-engine that drives them, having its long
history from Papin downwards ; there are the lathes in
which its cylinder was bored, and the string of ancestral
lathes from which those lathes proceeded ; there is the
steam-hammer under which its crank shaft was welded;
there are the puddling-furnaces, the blast-furnaces, the coal-
mines and the iron-mines needful for producing the raw
material ; there are the slowly improved appliances by
which the factory was built, and lighted, and ventilated;
there are the printing engine, and the die house, and the colour laboratory with its stock of materials from all parts of

the world, implying cochineal-culture, logwood-cutting, in-
digo-growing ; there are the implements used by the pro-
ducers of cotton, the gins by which it is cleaned, the elaborate machines by which it is spun : there are the vessels
in which cotton is imported, with the building-slips, the
rope-yards, the sail-cloth factories, the anchor-forges, needful for making them ; and besides all these directly necessary antecedents, each of them involving many others,
there are the institutions which have developed the requisite intelligence, the printing and publishing arrangements
which have spread the necessary information, the social organization which has rendered possible such a complex co-
operation of agencies.

Further analysis would show that the many arts thus
concerned in the economical production of a child's frock,
have each of them been brought to its present efficiency
by slow steps which the other arts have aided ; and that
from the beginning this reciprocity has been ever on the
increase. It needs but on the one hand to consider how
utterly impossible it is for the savage, even with ore and
coal ready, to produce so simple a thing as an iron hatchet ;
and then to consider, on the other hand, that it would have
been impracticable among ourselves, even a century ago,
to raise the tubes of the Britannia bridge from lack of the
hydraulic press ; to at once see how mutually dependent
are the arts, and how all must advance that each may advance. Well, the sciences are involved with each other
in just the same manner. They are, in fact, inextricably
woven into this same complex web of the arts ; and are
only conventionally independent of it. Originally the two
were one. How to fix the religious festivals ; when to sow :
how to weigh commodities ; and in what manner to measure ground ; were the purely practical questions out of
which arose astronomy, mechanics, geometry. Since then
there has been a perpetual inosculation of the sciences and

the arts. Science has been supplying art with truer generali
zations and more completely quantitative previsions. Art has
been supplying science with better materials and more perfect instruments. And all along the interdependence has been
growing closer, not only between art and science, but among
the arts themselves, and among the sciences themselves.

How completely the analogy holds throughout, becomes
yet clearer when we recognise the fact that *the sciences are
arts to each other.* If, as occurs in almost every case, the
fact to be analyzed by any science, has first to be prepared
—to be disentangled from disturbing facts by the afore
discovered methods of other sciences; the other sciences
so used, stand in the position of arts. If, in solving a dynamical problem, a parallelogram is drawn, of which the sides
and diagonal represent forces, and by putting magnitudes
of extension for magnitudes of force a measurable relation
is established between quantities not else to be dealt with;
it may be fairly said that geometry plays towards mechanics much the same part that the fire of the founder plays
towards the metal he is going to cast. If, in analyzing the
phenomena of the coloured rings surrounding the point of
contact between two lenses, a Newton ascertains by calculation the amount of certain interposed spaces, far too minute for actual measurement; he employs the science of
number for essentially the same purpose as that for which
the watchmaker employs tools. If, before writing down
his observation on a star, the astronomer has to separate
from it all the errors resulting from atmospheric and optical
laws, it is manifest that the refraction-tables, and logarithm-
books, and formulæ, which he successively uses, serve him
much as retorts, and filters, and cupels serve the assayer
who wishes to separate the pure gold from all accompanying ingredients.

So close, indeed, is the relationship, that it is impossible to say where science begins and art ends. All the in-

struments of the natural philosopher are the products of
art; the adjusting one of them for use is an art; there is
art in making an observation with one of them; it requires
art properly to treat the facts ascertained; nay, even the
employing established generalizations to open the way to
new generalizations, may be considered as art. In each of
these cases previously organized knowledge becomes the
implement by which new knowledge is got at: and whether
that previously organized knowledge is embodied in a tangible apparatus or in a formula, matters not in so far as its
essential relation to the new knowledge is concerned. If,
as no one will deny, art is applied knowledge, then such
portion of a scientific investigation as consists of applied
knowledge is art. So that we may even say that as soon
as any prevision in science passes out of its originally passive state, and is employed for reaching other previsions,
it passes from theory into practice—becomes science in action—becomes art. And when we thus see how purely
conventional is the ordinary distinction, how impossible it
is to make any real separation—when we see not only that
science and art were originally one; that the arts have
perpetually assisted each other; that there has been a constant reciprocation of aid between the sciences and arts;
but that the sciences act as arts to each other, and that the
established part of each science becomes an art to the
growing part—when we recognize the closeness of these
associations, we shall the more clearly perceive that as the
connexion of the arts with each other has been ever becoming more intimate; as the help given by sciences to
arts and by arts to sciences, has been age by age increasing; so the interdependence of the sciences themselves has
been ever growing greater, their mutual relations more involved, their *consensus* more active.

In here ending our sketch of the Genesis of Science, we

are conscious of having done the subject but scant justice. Two difficulties have stood in our way: one, the having to touch on so many points in such small space; the other, the necessity of treating in serial arrangement a process which is not serial—a difficulty which must ever attend all attempts to delineate processes of development, whatever their special nature. Add to which, that to present in anything like completeness and proportion, even the outlines of so vast and complex a history, demands years of study. Nevertheless, we believe that the evidence which has been assigned suffices to substantiate the leading propositions with which we set out. Inquiry into the first stages of science confirms the conclusion which we drew from the analysis of science as now existing, that it is not distinct from common knowledge, but an outgrowth from it—an extension of the perception by means of the reason.

That which we further found by analysis to form the more specific characteristic of scientific previsions, as contrasted with the previsions of uncultured intelligence—their quantitativeness—we also see to have been the characteristic alike in the initial steps in science, and of all the steps succeeding them. The facts and admissions cited in disproof of the assertion that the sciences follow one another, both logically and historically, in the order of their decreasing generality, have been enforced by the sundry instances we have met with, in which the more general or abstract sciences have been advanced only at the instigation of the more special or concrete—instances serving to show that a more general science as much owes its progress to the presentation of new problems by a more special science, as the more special science owes its progress to the solutions which the more general science is thus led to attempt—instances therefore illustrating the position that scientific advance is as much from the special to the general as from the general to the special.

Quite in harmony with this position we find to be the admissions that the sciences are as branches of one trunk, and that they were at first cultivated simultaneously; and this harmony becomes the more marked on finding, as we have done, not only that the sciences have a common root, but that science in general has a common root with language, classification, reasoning, art; that throughout civilization these have advanced together, acting and reacting upon each other just as the separate sciences have done; and that thus the development of intelligence in all its divisions and subdivisions has conformed to this same law which we have shown that the sciences conform to. From all which we may perceive that the sciences can with no greater propriety be arranged in a succession, than language, classification, reasoning, art, and science, can be arranged in a succession; that, however needful a succession may be for the convenience of books and catalogues, it must be recognized merely as a convention; and that so far from its being the function of a philosophy of the sciences to establish a hierarchy, it is its function to show that the linear arrangements required for literary purposes, have none of them any basis either in Nature or History.

There is one further remark we must not omit—a remark touching the importance of the question that has been discussed. Unfortunately it commonly happens that topics of this abstract nature are slighted as of no practical moment; and, we doubt not, that many will think it of very little consequence what theory respecting the genesis of science may be entertained. But the value of truths is often great, in proportion as their generality is wide. Remote as they seem from practical application, the highest generalizations are not unfrequently the most potent in their effects, in virtue of their influence on all those subordinate generalizations which regulate practice. And it must be so here. Whenever established, a correct theory of the historical development of the sciences must have an immense effect upon education; and, through education, upon civilization. Greatly as we differ from him in other respects, we agree with M. Comte in the belief that, rightly conducted, the education of the individual must have a certain correspondence with the evolution of the race.

No one can contemplate the facts we have cited in illustration of the early stages of science, without recognising the *necessity* of the processes through which those stages were reached—a necessity which, in respect to the leading truths, may likewise be traced in all after stages. This necessity, originating in the very nature of the phenomena to be analyzed and the faculties to be employed, more or less fully applies to the mind of the child as to that of the savage. We say more or less fully, because the correspondence is not special but general only. Were the *environment* the same in both cases, the correspondence would be complete. But though the surrounding material out of which science is to be organized, is, in many cases, the same to the juvenile mind and the aboriginal mind, it is not so throughout; as, for instance, in the case of chemistry, the phenomena of which are accessible to the one, but were inaccessible to the other. Hence, in proportion as the environment differs, the course of evolution must differ. After admitting sundry exceptions, however, there remains a substantial parallelism; and, if so, it becomes of great moment to ascertain what really has been the process of scientific evolution. The establishment of an erroneous theory must be disastrous in its educational results; while the establishment of a true one must eventually be fertile in school-reforms and consequent social benefits.

WORKS OF HERBERT SPENCER,

PUBLISHED BY

D. APPLETON AND COMPANY.

SYSTEM OF PHILOSOPHY

I.—FIRST PRINCIPLES.

(New and Enlarged Edition.)

PART I.—THE UNKNOWABLE.
PART II.—LAWS OF THE KNOWABLE.
559 pages. Price, ... $2.50

II.—THE PRINCIPLES OF BIOLOGY.—VOL. I.

PART I.—THE DATA OF BIOLOGY.
PART II.—THE INDUCTIONS OF BIOLOGY.
PART III.—THE EVOLUTION OF LIFE.
475 pages. Price, ... $2.50

PRINCIPLES OF BIOLOGY.—VOL. II.

PART IV.—MORPHOLOGICAL DEVELOPMENT.
PART V.—PHYSIOLOGICAL DEVELOPMENT.
PART VI.—LAWS OF MULTIPLICATION.
565 pages. Price, ... $2.50

III.—THE PRINCIPLES OF PSYCHOLOGY.

PART I.—THE DATA OF PSYCHOLOGY. 144 pages. Price, - . $0.75
PART II.—THE INDUCTIONS OF PSYCHOLOGY. 146 pages. Price, . $0.75
PART III.—GENERAL SYNTHESIS. 100 pages. } Price, - . $1.00
PART IV.—SPECIAL SYNTHESIS. 112 pages.

MISCELLANEOUS.

I.—ILLUSTRATIONS OF UNIVERSAL PROGRESS.

THIRTEEN ARTICLES. 451 pages. Price, ... $2.50

II.—ESSAYS:

MORAL, POLITICAL, AND ÆSTHETIC.

TEN ESSAYS. 386 pages. Price, ... $2.50

III.—SOCIAL STATICS:

OR THE CONDITIONS ESSENTIAL TO HUMAN HAPPINESS SPECIFIED, AND THE FIRST OF THEM DEVELOPED.

523 pages. Price, ... $2.50

IV.—EDUCATION:

INTELLECTUAL, MORAL, AND PHYSICAL.

283 pages. Price, ... $1.25

V.—CLASSIFICATION OF THE SCIENCES.

50 pages. Price, ... - $0.25

VI.—SPONTANEOUS GENERATION, &c.

16 pages Price, ... $0.25

THE DESCENT OF MAN.

THE ORIGIN OF CIVILIZATION;

OR, THE

PRIMITIVE CONDITION OF MAN.

By SIR JOHN LUBBOCK, Bart., M. P., F. R. S.

380 Pages. Illustrated.

This interesting work is the fruit of many years' research by an accomplished naturalist, and one well trained in modern scientific methods, into the mental, moral, and social condition of the lowest savage races. The want of a work of this kind had long been felt, and, as scientific methods are being more and more applied to questions of humanity, there has been increasing need of a careful and authentic work describing the conditions of those tribes of men who are lowest in the scale of development.

"This interesting work—for it is intensely so in its aim, scope, and the ability of its author—treats of what the scientists denominate *anthropology*, or the natural history of the human species; the complete science of man, body and soul, including sex, temperament, race, civilization, etc."—*Providence Press.*

"A work which is most comprehensive in its aim, and most admirable in its execution. The patience and judgment bestowed on the book are everywhere apparent; the mere list of authorities quoted giving evidence of wide and impartial reading. The work, indeed, is not only a valuable one on account of the opinions it expresses, but it is also most serviceable as a book of reference. It offers an able and exhaustive table of a vast array of facts, which no single student could well obtain for himself, and it has not been made the vehicle for any special pleading on the part of the author."—*London Athenæum.*

"The book is no cursory and superficial review; it goes to the very heart of the subject, and embodies the results of all the later investigations. It is replete with curious and quaint information presented in a compact, luminous, and entertaining form."—*Albany Evening Journal.*

"The treatment of the subject is eminently practical, dealing more with fact than theory, or perhaps it will be more just to say, dealing only with theory amply sustained by fact."—*Detroit Free Press.*

"This interesting and valuable volume illustrates, to some extent, the way in which the modern scientific spirit manages to extract a considerable treasure from the chaff and refuse neglected or thrown aside by former inquirers."—*London Saturday Review.*

D. APPLETON & CO., Publishers.

LAY SERMONS,
ADDRESSES, AND REVIEWS,

By THOMAS HENRY HUXLEY.

Cloth, 12mo.   390 pages.   Price, $1.75

Tuis is the latest and most popular of the works of this in-trepid and accomplished English thinker. The American edition of the work is the latest, and contains, in addition to the English edition, Professor Huxley's recent masterly address on "Spontaneous Generation," delivered before the British Association for the Advancement of Science, of which he was president.

The following is from an able article in the *Independent:*

The "Lay Sermons, Addresses, and Reviews" is a book to be read by every one who would keep up with the advance of truth—as well by those who are hostile as those who are friendly to his conclusions. In it, scientific and philosophical topics are handled with consummate ability. It is remarkable for purity of style and power of expression. Nowhere, in any modern work, is the advancement of the pursuit of that natural knowledge, which is of vital importance to bodily and mental well-being, so ably handled.

Professor Huxley is undoubtedly the representative scientific man of the age. His reverence for the right and devotion to truth have established his leadership of modern scientific thought. He leads the beliefs and aspirations of the increasingly powerful body of the younger men of science. His ability for research is marvellous. There is possible no more equipoise of judgment than that to which he brings the phenomena of Nature. Besides, he is not a mere scientist. His is a popularized philosophy; social questions have been treated by his pen in a manner most masterly. In his popular addresses, embracing the widest range of topics, he treads on ground with which he seems thoroughly familiar.

There are those who hold the name of Professor Huxley as synonymous with irreverence and atheism. Plato's was so held, and Galileo's, and Descartes's, and Newton's, and Faraday's. There can be no greater mistake. No man has greater reverence for the Bible than Huxley. No one more acquaintance with the text of Scripture. He believes there is definite government of the universe; that pleasures and pains are distributed in accordance with law; and that the certain proportion of evil woven up in the life even of worms will help the man who thinks to bear his own share with courage.

In the estimate of Professor Huxley's future influence upon science, his youth and health form a large element. He has just passed his forty-fifth year. If God spare his life, truth can hardly fail to be the gainer from a mind that is stored with knowledge of the laws of the Creator's operations, and that has learned to love all beauty and hate all vileness of Nature and art.

THE PHILOSOPHY OF EVOLUTION.

By HERBERT SPENCER.

This great system of scientific thought, the most original and important mental undertaking of the age, to which Mr. Spencer has devoted his life, is now well advanced, the published volumes being: *First Principles*, *The Principles of Biology*, two volumes, and *The Principles of Psychology*, vol. i., which will be shortly printed.

This philosophical system differs from all its predecessors in being solidly based on the sciences of observation and induction ; in representing the order and course of Nature ; in bringing Nature and man, life, mind, and society, under one great law of action ; and in developing a method of thought which may serve for practical guidance in dealing with the affairs of life. That Mr. Spencer is the man for this great work will be evident from the following statements:

" The only complete and systematic statement of the doctrine of Evolution with which I am acquainted is that contained in Mr. Herbert Spencer's ' System of Philosophy ; ' a work which should be carefully studied by all who desire to know whither scientific thought is tending."—T. H. HUXLEY.

" Of all our thinkers, he is the one who has formed to himself the largest new scheme of a systematic philosophy."—Prof. MASSON.

" If any individual influence is visibly encroaching on Mills in this country, it is his."—*Ibid.*

" Mr. Spencer is one of the most vigorous as well as boldest thinkers that English speculation has yet produced."—JOHN STUART MILL.

" One of the acutest metaphysicians of modern times."—*Ibid.*

" One of our deepest thinkers."—Dr. JOSEPH D. HOOKER.

It is questionable if any thinker of finer calibre has appeared in our country."—GEORGE HENRY LEWES.

" He alone, of all British thinkers, has organized a philosophy."—*Ibid.*

" He is as keen an analyst as is known in the history of philosophy ; I do not except either Aristotle or Kant."—GEORGE RIPLEY.

" If we were to give our own judgment, we should say that, since Newton, there has not in England been a philosopher of more remarkable speculative and systematizing talent than (in spite of some errors and some narrowness) Mr. Herbert Spencer."—*London Saturday Review.*

" We cannot refrain from offering our tribute of respect to one who, whether for the extent of his positive knowledge, or for the profundity of his speculative insight, has already achieved a name second to none in the whole range of English philosophy, and whose works will worthily sustain the credit of English thought in the present generation."—*Westminster Review.*

ON

THE ORIGIN OF SPECIES

BY

Means of Natural Selection;

OR,

THE PRESERVATION OF FAVORED RACES

IN THE

STRUGGLE FOR LIFE.

BY

CHARLES DARWIN, A. M.

One Volume. 12mo. Cloth. $2.00.

—— • ◆ • ——

"His first point is to show that species are in many cases not well defined, and that the whole order of natural history seems to be in a state of mutation, by reason of constant variations. Thus even under domestication, important changes may be introduced by intercrossing, by selection of the best individuals for propagation, by crossing parents marked by however slight, but favorable peculiarities.

"His second point is what he terms the universal and necessary struggle for existence. This follows from the high geometrical ratio of increase common to all beings. If there were no catastrophes, any one of the existing species would be sufficiently numerous in a few thousand years to cover the whole earth, to the exclusion of everything else.

"His third point is to prove that this struggle is directed by the law of natural selection. Even the races of domestic animals may be constantly improved and modified by choosing the best individuals for propagation. Nature brings the same discipline to bear upon the whole domain of animal and vegetable life. She seizes at once upon any slight variation that is favorable, and perpetuates it; in the universal pressure, any variation that is injurious is immediately extinguished."